A DIGEST OF DIGESTION

A Digest of
DIGESTION

Horace W. Davenport, D.Sc.

Professor of Physiology
The University of Michigan

YEAR BOOK MEDICAL PUBLISHERS, INC.
35 East Wacker Drive / Chicago

Library of Congress Catalog Card Number: 74-19955
International Standard Book Number: 0-8151-2325-6

PREFACE

Students are not learning gastrointestinal physiology to the same extent that they are learning cardiovascular, renal and respiratory physiology, and one reason is that they do not have an adequate text.

I intended my *Physiology of the Digestive Tract*, also published by Year Book Medical Publishers, to be an introductory text, but it turned out to contain more detail than students need or can handle during the few days allotted to gastroenterology in most current curricula. In writing this book, I have tried again to meet the needs of the beginning student.

References are for advanced students and professors, and accordingly there are no references in this book. Students who wish to know more than is given here can begin by reading my other book and by going to the library.

H.W.D.

TABLE OF CONTENTS

gastrin release by acid; effects of gastrin; cholecystokinin-pancreozymin (CCK-PZ); control of CCK-PZ
release; effects of CCK-PZ; secretin; control of secretin release; effects of secretin; synthesis and release of gastrin by ectopedic cells; trophic action of
gastrin; a secretin-like hormone in "pancreatic cholera"

Two components of pancreatic exocrine secretion: aqueous and enzyme juices; composition of the aqueous component; stimuli for secretion of the aqueous component; synergism of secretin and CCK-PZ; composition of the enzyme component; major digestive actions of the enzymes

Regulation of acidity and osmotic pressure of duodenal contents; digestion and absorption in the duodenum; duodenal regulation of gastric emptying; "dumping" as a consequence of interruption of duodenal regulation of gastric emptying; segmentation; control of segmentation; peristalsis; the ileocecal sphincter; gastroileal and gastrocolic reflexes; inhibitory reflexes

Volumes handled during digestion of a meal; bidirectional fluxes of water and electrolytes across the intestinal mucosa; net transport as a resultant of two opposing fluxes; absorption of water, sodium, chloride, bicarbonate and potassium; absorption of iron and calcium; the volume and composition of fluid entering the colon; absorption and secretion by the colon; causes of diarrhea

Structure of dietary carbohydrates; hydrolysis of starch by amylases; oligosaccharidases of the intestinal brush border and their actions; active absorption of glucose and galactose; effects of gluten enteropathy upon absorption; absence of oligosaccharidases: lactase deficiency

Absorption of intact native protein and of vitamin B_{12} bound to intrinsic factor; digestion and absorption of endogenous protein: enzymes and desquamated cells; normal and abnormal leakage of plasma proteins into the gut; hydrolysis of proteins; absorption

amounts of fat in the stool: steatorrhea; summary of errors in fat digestion and absorption; absorption of short- and medium-chain triglycerides; absorption of cholesterol

Sources of intestinal gas: swallowing, fermentation, neutralization of acid by bicarbonate, and diffusion; normal and abnormal quantities of gas in the gut; eructation; movement of gas through the gut; separation of gas from liquid or solid contents of the gut; flatus; entrapped gas—the cause of floating of stools

Innervation of the colon; absorption and secretion by the colonic mucosa; haustration and haustral shuttling; multihaustral propulsion; peristalsis; normal and abnormal rates of movement of colonic contents; stimuli for defecation; internal and external anal sphincters and their role on continence and defecation; the defecation reflex and the mechanics of defecation; death while straining at stool; megacolon; psychosomatic factors; normal and abnormal frequency of defecation; constipation and its consequences

1. CHEWING AND THE SECRETION OF SALIVA ᶜᴬᴸᴸ

The amount that food is chewed depends upon the nature of the food and upon habit, and it has little effect upon the subsequent processes of digestion.

Although chewing movements can be voluntary, most chewing during a meal is a rhythmic reflex stimulated by pressure of food against the gums, teeth, hard palate and tongue. This pressure causes relaxation of the muscles holding the jaw closed against gravity, and the subsequent partial opening of the mouth decreases the strength of stimuli exerted by the food. Rebound contraction of the jaw muscles follows, and the cycle is repeated about once a second.

Most persons chew on one side of the mouth at a time.

When the contents of the mouth have been sufficiently divided, movement of the tip of the tongue separates a small bolus from the rest and brings it to the midline. Then, with the forepart of the tongue pressed firmly against the teeth and with the mouth closed, the tongue propels the bolus backward to the oral pharynx, where pressure of the bolus against receptor endings stimulates the act of swallowing.

Presence of food in the mouth also stimulates secretion of saliva from the three pairs of salivary glands and from numerous small glands in the buccal cavity. The rate of flow of saliva may be as high as 4 ml a minute. Saliva dissolves sapid substances and makes them available for taste, and it lubricates the food for swallowing. When food is not being eaten, saliva keeps the mouth wet. Decreased secretion occurring during dehydration makes the mouth dry and contributes to the sensation of thirst. Saliva facilitates speech.

Secretion of saliva is entirely under control of autonomic nerves which innervate the glands. There is no hormonal

1

control of salivary secretion. Although adrenergic sympathetic stimulation evokes some secretion from the submaxillary glands, the major stimuli for secretion by all glands come through parasympathetic fibers. These are cholinergic, and administration of atropine diminishes salivary secretion and makes the mouth dry.

Fig. 1-1.—The composition of human parotid saliva as a function of its rate of secretion. The concentrations of the ions in plasma are shown at the right-hand margin. (Adapted from Thaysen, J. H., Thorn, N. A., and Schwartz, I. L.: Am. J. Physiol. 178:155, 1954.)

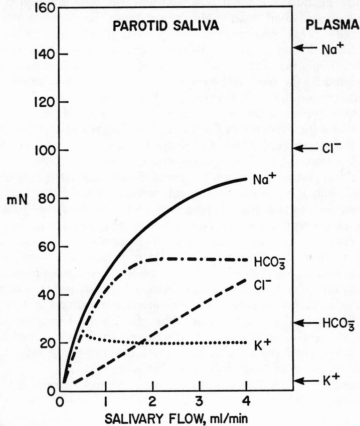

Human saliva is a hypotonic solution whose chief cations are Na^+ and K^+ (Fig. 1-1). In man, the potassium concentration at maximal rate of secretion is almost equal to that of plasma, and the sodium concentration is about three-fourths that of plasma. The anions are Cl^- and HCO_3^-, and the bicarbonate concentration is greater than that of plasma. Bicarbonate of saliva neutralizes acids of food, and because it neutralizes acids produced by bacteria in the mouth it helps to prevent caries.

Juice secreted by the submaxillary and sublingual glands contains mucin, which is responsible for much of the lubricating action of saliva. The juice of all three pairs of glands contains ptyalin, an amylase which catalyzes the hydrolysis of the α-1,4 glucosidic linkages of starch. The enzyme is stable between pH 4 and 11, and consequently it continues to act upon starch in the part of the gastric contents that has not become acid.

2. SWALLOWING

At rest, the esophagus is a flaccid, empty tube connecting the pharynx and the stomach. The pressure within it is the same as intrathoracic pressure (Fig. 2-1). Because intrathoracic pressure is below ambient pressure, there is a pressure gradient between the mouth and the esophagus. Air would enter the esophagus from the mouth were not the junction between the pharynx and the esophagus closed by the hypopharyngeal sphincter.

Intra-abdominal pressure is greater than atmospheric pressure and still greater than intrathoracic pressure. Consequently, there is a pressure gradient from stomach to esophagus. Regurgitation of gastric contents into the esophagus is prevented by the lower esophageal sphincter. When the esophagus is at rest, this sphincter maintains a zone of pressure greater than the pressure in the stomach or esophagus.

A person begins to swallow by detaching part of the contents of the mouth, either food, drink or saliva, with the tongue and thrusting it back into the pharynx. When the bolus touches the surface of the pharynx, it initiates impulses in afferent nerves going to the swallowing center in the medulla oblongata. The swallowing center responds by sending a rigidly ordered sequence of impulses to the muscles of the pharynx, the esophagus and the stomach.

Contraction of the striated muscles of the pharynx first closes the passage between the mouth and the nasopharynx to prevent regurgitation into the nose. The contractions generate a rapidly moving wave of high pressure, which thrusts the bolus into the pharynx, often accompanied by some air. As the bolus approaches the hypopharyngeal sphincter, muscles of the sphincter pull it open to allow the bolus to pass into the esophagus. Once the bolus has passed, the sphincter closes tightly. These events occur in less than a second.

5

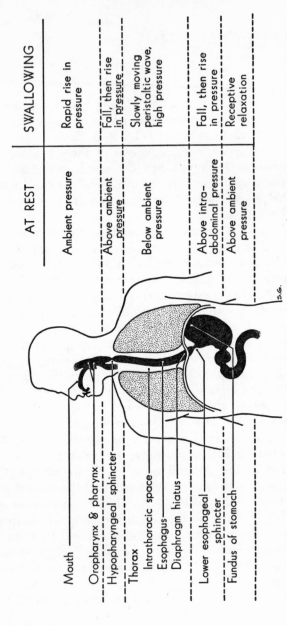

Fig. 2-1.—The pharynx, esophagus and stomach. Pressures at rest and during swallowing are shown.

As the bolus passes the entrance to the trachea, the epiglottis folds over the glottis, the glottis closes and respiration is briefly inhibited.

⌈Sequential, ringlike contractions of the muscle of the esophagus form a peristaltic wave which pushes the bolus toward the stomach.⌉Pressure generated by the peristaltic wave would also push the bolus back into the pharynx, but reflux is prevented by firm closure of the hypopharyngeal sphincter during the period in which high pressure exists in the upper esophagus. The peristaltic wave moves slowly down the esophagus, taking 5 to 9 seconds for the journey (Fig. 2-2).

When swallowing begins, the lower esophageal sphincter starts to relax, and it remains relaxed until the peristaltic wave at the lower end of the esophagus pushes the bolus through it into the stomach. The lower esophageal sphincter then contracts, and it maintains a pressure above the resting value for several seconds. In the meantime, the fundus of the stomach has partially relaxed to receive the bolus.

The muscles of the pharynx, hypopharyngeal sphincter and upper third of the esophagus are striated muscles innervated by motor neurons. The muscles are relaxed unless action potentials reach them through their nerves, and the vigor of their contraction is determined by the frequency of the impulses reaching them. They contract and relax in the pattern necessary to generate a moving wave of contraction, because the swallowing center sends out a stereotyped sequence of excitatory impulses to them.

The muscle of the lower two thirds of the esophagus and of the lower esophageal sphincter is smooth muscle, and its motor nerve is the vagus. Vagal cholinergic preganglionic fibers synapse with short postganglionic fibers embedded in the wall of the esophagus. Acetylcholine liberated by the postganglionic fibers causes the esophageal muscle to contract. During swallowing, the vagal fibers to the esophagus are activated in an orderly manner from above downward, so that peristalsis in the smooth-muscle segment of the esophagus follows smoothly the peristaltic wave in the striated muscle segment. Relaxation of the lower esophageal sphincter occurring at the end of esophageal peristalsis is also governed by the vagus, but the chemical mediator causing relaxation is neither acetylcholine nor norepinephrine.

When swallowing movements in the mouth follow one another rapidly, as in drinking, there are no peristaltic waves

Fig. 2-2.—Pressure changes occurring in the pharynx, esophagus and stomach during swallowing.

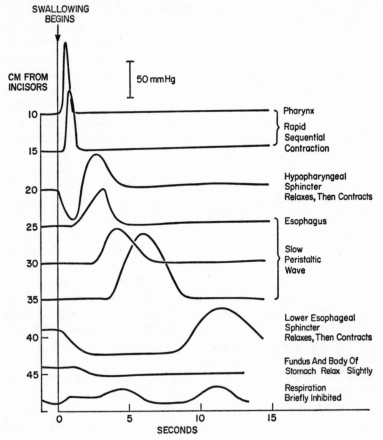

in the esophagus until after the last swallow; then a single normal peristaltic wave occurs.

An object falls because its specific gravity is greater than that of its surroundings. The average specific gravity of the thoracic contents is about 0.5. Consequently, in a man in the upright position, liquids, which are not held up by friction against the wall of the esophagus, fall quickly to the lower esophageal sphincter. There they wait about 5 seconds until the peristaltic wave arrives to push them through the sphincter into the stomach. On the other hand, the specific gravity of the abdominal contents is the same as that of food, and except for gas and the heavy x-ray contrast medium, barium sulfate, the position of the body has little influence upon the distribution of the contents of the digestive tract.

There are many receptors in the wall of the esophagus which send impulses along afferent nerves to the central nervous system. Some receptors respond to tension in the wall of the esophagus occurring during peristalsis, and the effect of their afferent impulses is to reinforce the outflow of the swallowing center and thereby to increase the strength of peristaltic contraction. If receptors in the wall of the esophagus are stimulated by some object in the lumen, for example, a piece of peanut butter sandwich stuck there, one or more peristaltic waves begin at the hypopharyngeal sphincter and move down the esophagus to the stomach. These waves occur without any preliminary movements of the mouth, and the subject is not aware that they are occurring. They are called *secondary peristaltic waves* to distinguish them from the *primary peristaltic waves* that follow swallowing movements of the mouth.

Receptors responding to tension in the esophageal wall send impulses that arouse the sensation of pain. The pain is referred to the anterior chest wall, sometimes to the posterior wall, the shoulder and the inside of the left arm. This distribution of referred pain is similar to that arising from myocardial ischemia, and it is sometimes difficult to distinguish between the esophagus and the heart as the source of the pain. Pain of variable nature and intensity, known as heart-burn, also occurs when the lower esophagus is inflamed by regurgitated acidic gastric contents.

Reflux of gastric contents is usually prevented by three mechanisms: mechanical, nervous and hormonal.

1. Pressure in the lower esophageal sphincter at rest is higher than pressure in the stomach or esophagus and therefore prevents flow of fluid from one to the other (Fig. 2-3). In normal adults, there is a short segment of the esophagus below the diaphragm, and this segment includes part of the lower esophageal sphincter. When intra-abdominal pressure rises, so does pressure in that part of the sphincter lying below the diaphragm, simply because intra-abdominal pressure is transmitted to that part of the sphincter as well as to the rest of the abdominal contents. This mechanism does not operate in infants who have no subdiaphragmatic segment of the esophagus. Heartburn occurring in women in midpregnancy is also attributed to absence of a subdiaphragmatic segment of the esophagus.

2. Pressure in the resting lower esophageal sphincter is controlled in part by afferent impulses in the vagus nerve, and these in turn are influenced by afferent impulses. When intra-abdominal pressure rises, pressure in the lower esophageal sphincter rises at the same time (Fig. 2-4), and the rise in pressure is the result of reflex activation of efferent fibers to the sphincter. In persons whose lower esophageal sphincter and upper part of the stomach are herniated through the hiatus of the diaphragm, the external pressure on the sphincter is intrathoracic pressure. Intrathoracic pressure is lower than intra-abdominal pressure. On account of the position of the sphincter in the thorax, the mechanical effect of an increase in intra-abdominal pressure upon the sphincter cannot operate. Nevertheless, in many persons with hiatal hernia, an increase in intra-abdominal pressure is accompanied by an increase in sphincter pressure sufficient to prevent reflux. The increase in sphincter pressure is the result of a vagally me-

diated reflex. In some persons, the increase in sphincter pressure accompanying an increase in intra-abdominal pressure is below normal, and those persons have frequent reflux of gastric contents into the esophagus (Fig. 2-4).

3. The hormone gastrin (Chap. 5), the major action of which is to stimulate secretion of acid by the gastric mucosa, also stimulates contraction of the lower esophageal sphincter.

Fig. 2-3.—The pressure barrier between the stomach and the esophagus is called the lower esophageal sphincter. A tube connected to a pressure-measuring transducer is swallowed until its pressure-sensing tip is in the stomach, and it is slowly withdrawn into the esophagus. Pressure in the stomach is above atmospheric pressure, but as the tube is withdrawn, a segment of higher pressure approximately 2 cm long is found. After the tip of the tube has passed the diaphragm and entered the esophagus, subatmospheric pressure is encountered. In actual practice, variations in pressure caused by respiratory movements are superimposed on the pressure tracing.

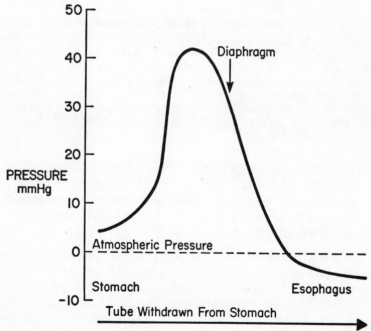

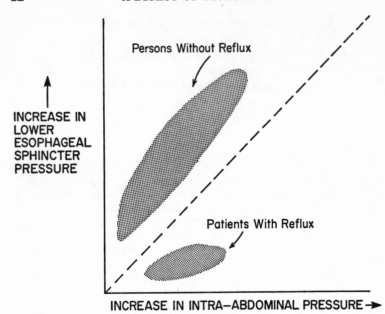

INCREASE IN LOWER ESOPHAGEAL SPHINCTER PRESSURE

INCREASE IN INTRA–ABDOMINAL PRESSURE →

Fig. 2-4.—As intra-abdominal pressure increases, pressure in the lower esophageal sphincter increases as well. In normal persons, the increase in sphincter pressure is greater than the increase in intra-abdominal pressure, and no reflux of gastric contents into the esophagus occurs. In patients with reflux, the increase in sphincter pressure is less than the increase in intra-abdominal pressure. (Adapted from Cohen, S., and Harris, L. D.: New Engl. J. Med. 284:1053, 1971.)

The concentration of gastrin in plasma rises during a meal, when the stomach is full and acid is being secreted. This rise in plasma gastrin may help to increase the competence of the sphincter at a time when sphincter competence is most important.⌉

In some persons, the lower esophageal sphincter fails to relax during swallowing; instead, it contracts more vigorously. Failure to relax is called *achalasia*. Passage of food from esophagus to stomach is greatly delayed, and the esophagus may become enormously dilated. Because a meal remains in the esophagus for a long time, food may be aspirated into the trachea, and persons with achalasia are prone to aspiration pneumonia. The underlying defect seems to be destruction of parasympathetic postganglionic fibers within the wall of the esophagus. Destruction of nerves innervating smooth-muscle cells makes the cells highly sensitive to the mediator that had been released by the nerves before they were destroyed (Cannon's law of denervation). Because the preganglionic fibers that liberate acetylcholine terminate within the wall of the esophagus, muscle cells that have become exquisitely sensitive to acetylcholine are exposed to the chemical mediator whenever the preganglionic fibers are reflexly activated. Consequently, the affected part of the esophagus goes into spasm. Sensitivity of the cells is used for differential diagnosis. A dose of a cholinergic drug that has no effect upon a normal person causes the esophagus of a patient with achalasia to contract strongly and painfully.

3. VOMITING

Retching and vomiting are governed by a center in the medulla oblongata which receives afferent information from two classes of receptors: (1) those in the viscera, particularly in the duodenum, which respond to constituents of the chyme such as ingested ipecac or copper salts, and (2) those in the chemoreceptor trigger zone in the area postrema of the brainstem which respond to blood-borne compounds produced in radiation sickness, in many diseases and to injected emetics such as apomorphine. In addition, vomiting is aroused by many events affecting the central nervous system. Some persons can vomit at will, and others vomit at the thought of nauseating objects. The sensation of nausea and the act of vomiting also occur in motion sickness and follow injuries producing crushing pain.

The complete act of vomiting begins with copious salivation. The antrum of the stomach contracts with great vigor, and there may be reverse peristalsis. The duodenum goes into spasm, and although it may not have reverse peristalsis its bile-stained contents are forced into the stomach. On the other hand, the body of the stomach and the lower esophageal sphincter relax completely (Fig. 3-1).

Slow, deep inspirations against a partially closed glottis produce moaning sounds and reduce intrathoracic pressure far below atmospheric pressure. At the same time, strong contractions of the abdominal muscles increase intra-abdominal pressure. The large pressure gradient from abdomen to thorax forces contents of the flaccid body of the stomach through the relaxed lower esophageal sphincter into the esophagus. Reverse peristalsis never occurs in the human esophagus. The hypopharyngeal sphincter remains closed, and no gastric contents enter the mouth. Distention of the

15

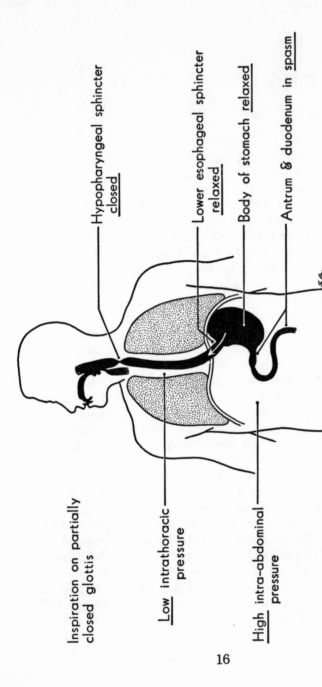

Inspiration on partially
closed glottis

Hypopharyngeal sphincter
closed

Lower esophageal sphincter
relaxed

Body of stomach relaxed

Low intrathoracic
pressure

High intra-abdominal
pressure

Antrum & duodenum in spasm

Fig. 3-1.—The mechanics of retching. Duodenal contents are forced into the stomach by spasm of the duodenum. The antrum is also in spasm, but the body of the stomach and the lower esophageal sphincter are relaxed. Inspiration against a partially closed glottis lowers intrathoracic pressure, and contraction of abdominal muscles raises intra-abdominal pressure. The pressure gradient from abdomen to thorax forces contents of the body of the stomach into the esophagus. The hypopharyngeal sphincter remains closed, and secondary peristaltic waves in the esophagus force its contents back into the stomach. The cycle may be repeated many times.

16

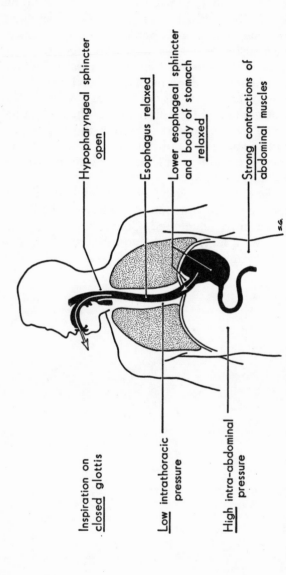

Inspiration on
closed glottis

Low intrathoracic
pressure

High intra-abdominal
pressure

Hypopharyngeal sphincter
open

Esophagus relaxed

Lower esophageal sphincter
and body of stomach
relaxed

Strong contractions of
abdominal muscles

Fig. 3-2.—The mechanics of vomiting. Antrum and duodenum are in spasm, but the body of the stomach and the lower esophageal sphincter are relaxed. Strong inspiration with a closed glottis lowers intrathoracic pressure, and strong contractions of abdominal muscles force the contents of the stomach into the esophagus and through the relaxed hypopharyngeal sphincter. There may be reverse peristalsis in the antrum of the stomach, but there is no reverse peristalsis in the body of the stomach or in the esophagus.

esophagus induces secondary peristalsis, and gastric contents are swept back into the stomach. This is the act of retching, and it may be repeated many times before vomiting occurs.

At the end of a series of heaves, a person about to vomit thrusts out his jaw and opens his hypopharyngeal sphincter. At the same time, he takes a deep inspiration against a closed glottis and strongly contracts his abdominal muscles. The pressure gradient generated by contraction of the abdominal muscles forces the gastric contents through the esophagus, the hypopharyngeal sphincter and the mouth (Fig. 3-2). The body of the stomach remains flaccid and does not contribute to the expulsion of its contents.

In a young infant, there may be no intra-abdominal segment of the esophagus, and function of the lower esophageal sphincter is not yet fully developed. Consequently, regurgitation occurring after a large meal is more like the bubbling over of a fumarole than vomiting.

4. MUSCLE AND NERVES OF THE GUT

Smooth muscle of the gut is different in structure and function from striated muscle. The two kinds of muscle are contrasted in Table 4-1.

Individual smooth-muscle fibers are very small, and they are organized in bundles of about 200 cells in cross-section. The cells are connected one with the other by gap junctions through which current can flow from one cell to the other. The result is that each bundle is in effect a physiologic syncytium. The bundles are arranged in layers. A thin longitudinal layer runs lengthwise along the second half of the esophagus, the whole of the stomach and the small intestine. Strips of the longitudinal layer form the taeniae coli of the colon. Beneath the longitudinal layer is a thicker and stronger circular layer of bundles, each bundle roughly forming a ring around its hollow organ. Another thin layer of muscle, the muscularis mucosae, lies between the circular muscle and the mucosa.

The muscles of the pharynx, the hypopharynx and the first third of the esophagus are striated muscles, and they are totally dependent upon their efferent nerves for any function. At the other end of the gut, the external anal sphincter is also composed of striated muscle, and if its efferent nerves are destroyed, the sphincter is paralyzed. Between these two extremes lie the smooth-muscle structures of the gut, and the most important point to understand about the function of these structures is that they are provided with a nervous system of their own which is capable of executing the functions of the gut without any extrinsic innervation whatever. The extrinsic innervation this nervous system receives from the sympathetic and parasympathetic nerves modulates but does not command its activity.

19

TABLE 4-1.   COMPARISON OF SMOOTH MUSCLE OF THE GUT WITH
STRIATED MUSCLE

SMOOTH MUSCLE OF THE GUT	STRIATED MUSCLE
Very small fibers organized in bundles	Long fibers organized in muscle units
Muscle fibers within bundles connected by gap junctions	Fibers independent of each other except by common innervation by one motor neuron
Origin and insertion in connective tissue	Origin and insertion on bones
Innervated by fibers of nervous plexuses; a potentially independent nervous system that can act without the CNS	Innervated by motor neurons; entirely dependent on the CNS
Muscle fibers have no motor end plates; efferent fibers from plexuses liberate mediators near the cell surface	Each fiber has a motor end plate
Ganglion cells of plexuses innervated by preganglionic parasympathetic fibers and equivalent to parasympathetic postganglionic cell bodies	No plexuses
Ganglion cells of plexuses innervated by postganglionic sympathetic fibers; sympathetic activity modulates activity of ganglion cells by inhibiting on-going activity and reflexes	
Sympathetic postganglionic innervation of muscle cells of minor importance; but circulating sympathetic mediators have major influence on muscle cells	Circulating sympathetic mediators have minor effects on muscle cells
Muscle fiber membrane potential variable at rest	Membrane potential constant at rest
Spontaneous variations in membrane potential	None
Hyperpolarization with increased threshold	
Hypopolarization with decreased threshold	
Basic electrical rhythm = rhythmic depolarization conducted along longitudinal fiber bundles	

Smooth Muscle of the Gut	Striated Muscle
Electrotonic current flow between longitudinal fiber bundles and circular fiber bundles, influencing the threshold of circular fiber bundles	None
Action potentials may or may not occur during rhythmic depolarization	Action potentials follow end plate depolarization

Occurrence of action potentials strongly influenced by
    Activity in plexuses: cholinergic output decreasing muscle fiber threshold
    Circulating hormones: epinephrine and norepinephrine increasing muscle fiber threshold; gastrin and secretin with specific effects on specific muscles
	No plexuses; slight effect of circulating hormones

Active tension always present; modulated by mediators and hormones; varies with resting membrane potential	No active tension without muscle action potential
Variable response to stretch: elastic lengthening plus either stress relaxation or active contraction	Elastic lengthening only
Variable response to release of stretch: elastic shortening plus either active contraction or further relaxation	Elastic shortening only
Active tension increases following action potential	No active tension without muscle action potential
Very wide range of active tension; large tetanus-to-twitch ratio	Narrow range of tetanus-to-twitch ratio; large range of tension achieved by spatial and temporal summation
Slow contraction and relaxation	Rapid contraction and relaxation

Each smooth-muscle fiber has a transmembrane potential, positive outward and negative inward. This membrane potential, like that of striated muscle and nerve, is in part a potassium-diffusion potential. The concentration of potassium is higher inside the smooth-muscle cell than it is in the extracellular fluid, and the tendency of potassium ions to diffuse outward along their concentration gradient, carrying with them a positive charge, contributes to the transmembrane potential. Other ions also contribute to the transmembrane potential of smooth-muscle cells; both sodium and calcium ions tend to diffuse inward, carrying positive charges which oppose those of potassium diffusing in the opposite direction. However, the transmembrane potential of intestinal smooth-muscle cells cannot be explained entirely as the resultant of several diffusion potentials. The muscle cell membrane appears to contain an electrogenic sodium pump. This pump, using metabolic energy, pumps sodium ions out of the cell and therefore tends to make the outside of the cell positive with respect to the inside.

Variations in the activity of the electrogenic sodium pump account for much of the variability of the resting membrane

Fig. 4-1.—The effect of epinephrine upon a smooth-muscle cell of the longitudinal and circular muscle bundles of the gut. Epinephrine hyperpolarizes the cell. When the transmembrane potential is greater than threshold, action potentials do not occur. A decrease in tension follows the decrease in frequency of action potentials and the increase in transmembrane potential. Norepinephrine has the same effect.

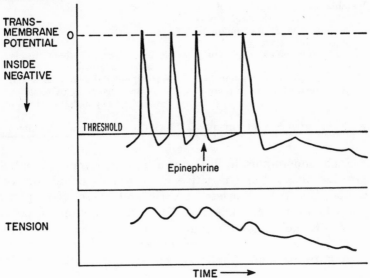

potential of smooth-muscle cells. The pump is stimulated by circulating epinephrine and norepinephrine, and as it vigorously extrudes sodium ions, the transmembrane potential becomes larger. The cells are hyperpolarized and less excitable, and their tension decreases (Fig. 4-1). On the other hand, acetylcholine depolarizes the cells, decreasing their membrane potential, increasing their tension and making them more excitable (Fig. 4-2).

When a smooth-muscle cell is depolarized to a certain extent, its membrane has an action potential. Like the action potentials of striated muscle cells and neurons, this is an abrupt depolarization which spreads over the whole of the cell's membrane and which is followed by repolarization.

Fig. 4-2.—The effect of acetylcholine upon a smooth-muscle cell of the longitudinal and circular muscle bundles of the gut. Acetylcholine causes depolarization of the cell, and when the membrane potential reaches threshold, an action potential occurs. Since acetylcholine increases the rate of depolarization, it increases the frequency of action potentials. An increase in tension follows the action potentials.

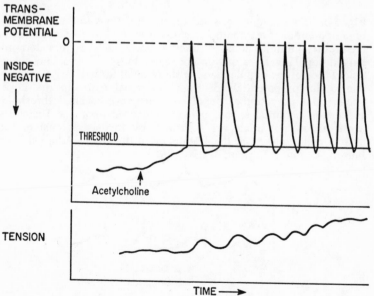

Smooth-muscle cells have three kinds of variations in their resting membrane potential (Fig. 4-3).

1. The membrane potential may slowly increase, in which case the cell is hyperpolarized and less excitable, or the membrane potential may slowly decrease, in which case the cell is hypopolarized and more excitable. These changes may be spontaneous in the sense that their cause is not identified, or they may be the effects of changes in membrane permeability and sodium-pumping resulting from changes in concentration of hormones or chemical mediators in the cell's environment.

2. Some smooth-muscle cells of the gut may have rhythmic depolarizations similar to the prepotentials of the sino-atrial node of the heart. If one of these depolarizations reaches threshold, an action potential follows. The frequency of action potentials in a particular cell, therefore, depends upon (a) the frequency with which the prepotentials occur, (b) the rate at which the prepotentials move to threshold and (c) the threshold at the moment.

Fig. 4-3.—Two samples of spontaneous variations in the transmembrane potential of a smooth-muscle cell in the longitudinal layer of the rabbit jejunum. The upper tracing shows the BER, repeated prolonged depolarizations that do not reach threshold for action potentials. Three of the depolarizations show small further variations similar to the prepotentials of sinoatrial nodal cells. In the lower tracing, the prepotential depolarizations have reached threshold, and action potential spikes are superimposed upon the BER. Because action potentials are followed by increased tension, the muscle cell contracts at the frequency of the BER. (Adapted from Bortoff, A.: Am. J. Physiol. 201:203, 1961.)

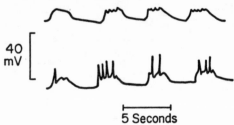

40
mV

5 Seconds

3. Some cells of the longitudinal muscle bundles (but not those of the circular muscle bundles) have spontaneous rhythmic depolarizations occurring at a frequency characteristic of the organ. In the stomach, these cells are located high on the greater curvature, and their frequency of depolarization is approximately three depolarizations per minute. This train of spontaneous depolarizations, called the Basic Electrical Rhythm (BER) or Pacesetter Potential (PSP), is conducted along the longitudinal muscle bundles, and in the stomach it is responsible for establishing the frequency and the rate of progress of peristaltic waves.

Smooth-muscle cells contain myofilaments of actin and myosin, but the filaments are not arranged in the elegant array characteristic of striated muscle fibers. The intimate mechanism by which chemical energy is transformed into mechanical energy appears to be the same in both kinds of muscle. The chemical machine in smooth-muscle cells, however, is continuously active; smooth-muscle cells of the gut always have some degree of tension during life. This degree of tension is related to the membrane potential: a decrease in transmembrane potential, or hypopolarization, is associated with an increase in active tension, and an increase in transmembrane potential, or hyperpolarization, is associated with a decrease in tension, or relaxation. In addition, an action potential is followed by an increase in active tension. Changes in active tension are slow, and fusion of contractions following action potentials occurs readily. Tetanus of smooth muscle occurs at a low frequency of action potentials. The range of tension development by gut smooth muscle, from almost complete relaxation to maximal contraction, is very wide.

An action potential is conducted over the membrane of a single cell, but it cannot be conducted from a single cell to another single adjacent cell. When an action potential occurs in a cell's membrane, the transmembrane potential becomes zero, or it overshoots and reverses its sign. Consequently, there is a potential difference between the depolarized cell and its neighboring, undepolarized cell, and current flows from the positively charged surface of the undepolarized cell through the extracellular fluid to the depolarized cell. The circuit is completed by flow of current from the interior of the depolarized cell to the interior of the adjacent cell through the gap junction connecting the two cells. If the amount of current flowing is sufficient to depolarize the neighboring cell to threshold, an action potential will occur in the neighboring cell, and the action potential will, in effect, be conducted from one cell to another. The difficulty in smooth muscle of the gut is that each cell is very small, and therefore the surface-to-volume ratio is very large. A completely depolarized cell acts as a current sink with respect to its neighbors, but because the volume of the cell is small, not enough current flows to bring the membrane of the neighboring cells to threshold (Fig. 4-4). To circumvent this problem, smooth-muscle cells in the gut are organized in bundles. If a thousand cells, a group five cells long and two hundred cells in cross-section, are simultaneously depolarized, they form a large current sink. Enough current flows from the neighbors of these thousand cells so that the neighbors themselves reach threshold and experience action potentials. Excitation is thus conducted along bundles, and the cells of the bundles contract as a unit.

As the conducted wave of partial depolarization known as the BER passes slowly along bundles of the longitudinal muscle, it acts as a current sink for the underlying circular muscle. Therefore, there is an electrotonic flow of current from successive rings of circular muscle to the longitudinal muscle. This current flow tends to depolarize the bundles of circular muscle and bring them toward their threshold for action potentials. Whether or not the circular muscle bundles reach threshold, have action potentials and contract depends on their threshold, or excitability, at the moment, and their threshold at the moment depends in turn upon (1) the amount of circulating epinephrine and norepinephrine and (2) the amount of acetylcholine liberated near the muscle fiber bundles by nerves of the local plexuses.

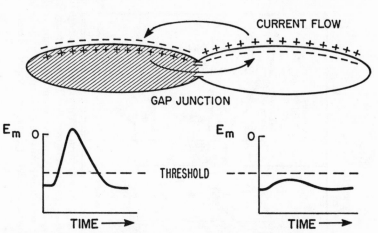

Fig. 4-4.—Current flow between a depolarized smooth-muscle cell of the gut and an adjacent polarized cell. Current flows in the external circuit from the surface of the polarized cell to the depolarized cell, here represented as having a reversal or overshoot of its membrane potential. The circuit is completed by flow of current from the depolarized cell to its neighbor through their gap junction. The cells are so small that current flow is not enough to bring the polarized cell to threshold and to initiate an action potential. When about 1,000 smooth-muscle cells in a bundle are simultaneously depolarized, however, current flow is sufficient to bring neighboring cells to threshold, and action potentials are propagated along the bundle.

There are two major networks of nerve fibers forming the plexuses of the gut: (1) the myenteric plexus between the two muscle layers and (2) the submucous plexus lying between the mucosa and the circular muscle layer. Together these form a complete and competent nervous system, and they are the evolutionary descendants of the primitive nervous system in the wall of such animals as the sea anemone.

The plexuses contain cells whose afferent fibers end in receptors in the wall of the gut or in the mucosa (Fig. 4-5). The receptors may be chemoreceptors sensitive to components of the chyme such as hydrogen ions or polypeptides, or

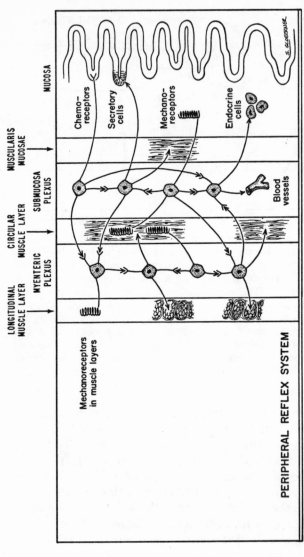

Fig. 4-5.—The peripheral reflex system of the gut. The intrinsic plexuses form a nervous system capable of integrative action without the aid of the extrinsic nerves. Chemo- and mechanoreceptors send impulses to cell bodies of the plexuses. The cell bodies have efferent processes that synapse with many other cells of the plexuses, and efferent fibers from the latter cells stimulate or inhibit muscle cells in the muscle layers and in the blood vessels, stimulate secretory cells of the mucosa or regulate release of hormones from endocrine cells.

they may be mechanoreceptors sensitive to stretch or tension. Efferent fibers of the receptor cells synapse with cell processes and cell bodies within the plexuses to form an integrating nervous system. Efferent fibers leave cells of the plexuses to reach muscle cells of the longitudinal and circular layers and of the muscularis mucosae. When the efferent fibers liberate acetylcholine, they are excitatory. By liberating other mediators, they may be inhibitory. Efferent fibers also reach secretory cells of the mucosa, and their mediators stimulate secretion of juices into the lumen or hormones into the blood. Distention of the stomach by neutral protein digestion products provides an example of the action of this intrinsic nervous system. Distention activates stretch receptors and, through a cholinergic reflex containing at least one synapse, stimulates release of gastrin by cells in the antrum and secretion of acid by oxyntic cells in the body of the stomach. The polypeptides stimulate chemoreceptors and, through a similar cholinergic reflex, they too stimulate release of gastrin and secretion of acid.

Extrinsic innervation arrives by two pathways: (1) parasympathetic preganglionic fibers running in the vagus nerve (Fig. 4-6) and (2) sympathetic postganglionic fibers running along blood vessels (Fig. 4-7).

Parasympathetic preganglionic fibers end on ganglion cells of the intrinsic plexuses of the gut, and therefore the fibers of the ganglion cells can be called postganglionic fibers. Such nomenclature is misleading and meaningless, however. In all parasympathetic systems outside the gut, the postganglionic fibers have no function except that dictated by their preganglionic innervation. In the gut, the ganglion cells and their processes are part of an autonomous nervous system capable of acting without parasympathetic innervation. The function of the parasympathetic innervation is to modulate, not absolutely control, the activity of the nervous system contained in the plexuses.

Most of the parasympathetic fibers to the plexuses of the gut are cholinergic and excitatory. An example of their action is seen during hunger. Then the glucose-sensitive cells of the

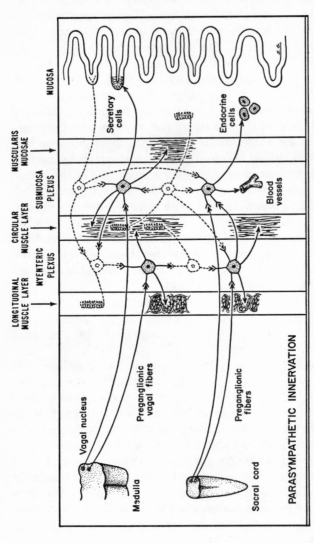

Fig. 4-6.—The parasympathetic innervation of the gut. Parasympathetic preganglionic fibers synapse with cells of the intrinsic plexuses and regulate the activity of the plexuses. In many instances, the parasympathetic fibers are cholinergic and excitatory, and their action increases motility, secretion and release of hormones. In other important instances, they are noncholinergic and inhibitory, and they mediate such reflexes as relaxation of the lower esophageal sphincter and the stomach during swallowing.

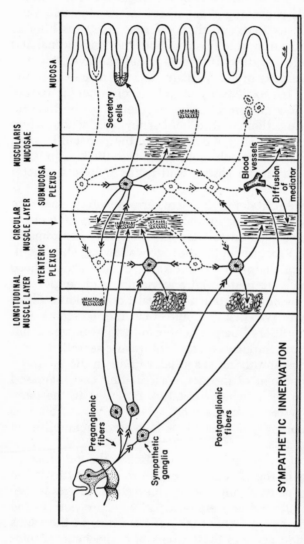

Fig. 4-7.—The sympathetic innervation of the gut. Postganglionic sympathetic fibers synapse with the cells of the intrinsic plexuses and regulate the activity of the plexuses. In most instances, they are adrenergic and inhibitory, and their action is to inhibit on-going activity in the plexuses. Many sympathetic adrenergic fibers innervate the smooth muscle of the blood vessels, causing vasoconstriction, and norepinephrine released at the blood vessels inhibits nearby cells of the muscle layers. Some adrenergic sympathetic fibers end near cells of the muscle layers. Action of the sympathetic nerves to the gut is backed up by epinephrine and norepinephrine released by the adrenal medulla.

31

hypothalamus excite the cells of the vagal nucleus, and action potentials run along vagal fibers to the stomach. Ganglion cells are excited, and their efferent processes liberate acetylcholine near muscle fibers of both the longitudinal and circular muscle bundles. In the longitudinal layer, the frequency and velocity of the BER are slightly increased. In the circular layer, the excitability of the muscle fibers is raised; electrotonic flow of current between them and the longitudinal layer brings them to threshold. They have action potentials and contract. A ring of contraction, the peristaltic wave, sweeps over the stomach, following the BER as it traverses from high on the gastric wall to the pyloric sphincter (Fig. 4-8).

Not all efferent fibers in the vagus are excitatory; some are inhibitory. An example is furnished by the innervation of the lower esophageal sphincter and the body of the stomach. Both of these structures relax during swallowing, and the efferent fibers mediating relaxation are in the vagus. These particular fibers are neither adrenergic nor cholinergic. The best present evidence is that the vagal preganglionic fibers liberate 5-hydroxytryptamine and that the corresponding postganglionic fibers liberate adenosine triphosphate. Other vagal fibers carry impulses that inhibit gastric secretion.

There are apparently no vasodilator fibers in the parasympathetic innervation of the gut. Vasodilation and increased blood flow do occur when excitatory impulses in the parasympathetic system enhance motility and secretion, but this is active hyperemia, the result of increased metabolism of muscle and glands.

The fact that parasympathetic impulses to the gut enhance reflex activity of the plexuses does not mean that this is the only function of parasympathetic innervation. There are reflexes whose afferent and efferent pathways are both in the vagal or sacral nerves. About 90% of the fibers in the vagus are afferent, and many of these fibers have mechanoreceptor or chemoreceptor endings in the stomach, intestine and proximal colon. Receptor endings in the distal colon send impulses to the sacral section of the spinal cord and, among

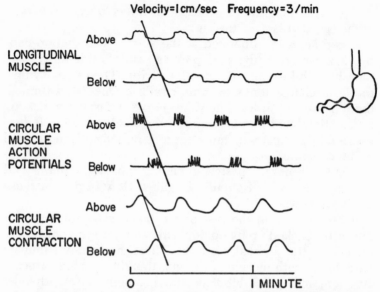

Fig. 4-8.—The origin and control of peristaltic waves in the antrum of the stomach. A pair of electrodes, 6.8 cm apart, in the longitudinal muscle record the BER at a frequency of 3 per minute and moving from the body of the stomach toward the pyloric sphincter at a velocity of 1 cm per second. Action potentials in the circular muscle accompany the BER, and consequently a wave of peristaltic contraction sweeps over the antrum at the velocity of the BER. Two or more waves can often be seen in sequence in the antrum of the human stomach.

other things, enhance the defecation reflex. An example of a vagally mediated reflex is the relaxation of the stomach that follows distention of the esophagus. Impulses arising in stretch receptors in the esophagus travel centrally in the vagus, and efferent impulses travel in nonadrenergic, noncholinergic vagal fibers to the stomach.

The sympathetic nervous system affects the gut through four pathways: (1) by adrenergic postganglionic fibers terminating on ganglion cells of the plexuses, (2) by adrenergic postganglionic fibers going to some muscle bundles, (3) by epinephrine and norepinephrine from the adrenal medulla, and (4) by adrenergic vasoconstrictor fibers to smooth muscle of the blood vessels of the gut.

Norepinephrine liberated at the surface of ganglion cells by sympathetic fibers depresses their excitability and thereby inhibits any on-going activity. Therefore, sympathetic discharge tends to inhibit reflexes mediated through the plexuses or through the vagus nerve. Thus, the peristaltic reflex stimulated by distention of the small intestine is suppressed by sympathetic impulses to ganglion cells in the wall of the intestine.

Some adrenergic postganglionic sympathetic fibers end near muscle cells of circular bundles. How important these are is unknown.

In the gut, as in the rest of the body, sympathetic influences are backed up by epinephrine and norepinephrine liberated by the adrenal medulla. These hormones stimulate vascular smooth muscle of the gut, thereby causing vasoconstriction, and they inhibit intestinal smooth muscle, thereby decreasing motility.

Adrenergic postganglionic sympathetic fibers innervate the smooth muscle of the blood vessels of the gut. When they are stimulated, there is an immediate vasoconstriction and eventually decreased intestinal motility. This decrease in motility is probably the result of diffusion of norepinephrine from the blood vessels to neighboring smooth-muscle cells of the longitudinal and circular bundles.

5. HORMONES OF THE GUT

The three best understood hormones of the gut are gastrin, cholecystokinin-pancreozymin (CCK-PZ) and secretin. All three are polypeptides. Their structures are given in Table 5-1, and their major actions are listed in Table 5-2. The structures need not be memorized, but their homologies should be recognized.

Gastrin is a straight chain of 17 amino acids, a heptadecapeptide. The N-terminal amino acid is pyroglutamic acid, and the carboxyl group of the C-terminal phenylalanine is aminated. Gastrin has several variants. In one, the tyrosine at position 12 is sulfated, but whether or not the tyrosine is sulfated makes little difference in the action of the hormone. In addition to these two gastrins, there is a "Big Gastrin" having a molecular weight of about 4,000 and consisting of the heptadecapeptide bound to a larger molecule. At least three other forms are found circulating in plasma: a still bigger "Big-Big Gastrin" and two "Mini-Gastrins," each consisting of thirteen amino acids. Each circulating gastrin has a different half-life, and the gastrins differ from one another in potency. The relative importance of the various forms of circulating gastrin has not been determined.

In all gastrins, the last four amino acids, counting from the N-terminus, are -Trp-Met-Asp-Phe-NH$_2$, and these four amino acids constitute the active group of the molecules. A tetrapeptide of this sequence has all the physiologic actions of gastrin, although on a molar basis it is less potent than the larger molecules. The synthetic compound, pentagastrin, consists of these four amino acids linked to a substituted β-alanine, and it is replacing histamine and its analogs in tests of acid secretion. The longer chain of amino acids attached to the four amino acids of the active group confer on gastrin its

TABLE 5-1. MAJOR HORMONES OF THE GUT

HUMAN GASTRIN II	ACTIVE GROUPS

1 2 3 4 5 6 7 8 9 10 11 12 13 | 14 15 16 17
Pyr-Gly-Pro-Trp-Leu-Glu-Glu-Glu-Glu-Glu-Ala-Tyr-Gly- | -Trp-Met-Asp-Phe-NH_2
$\qquad\qquad\qquad\qquad\qquad\qquad\qquad\qquad$ SO_3

Human gastrin I is not sulfated at Tyr_{12}

CHOLECYSTOKININ-PANCREOZYMIN

$\qquad\qquad\qquad\qquad\qquad\qquad$ 26 27 28 29 | 30 31 32 33
$\qquad\qquad\qquad\qquad$ 25 more-Asp-Tyr-Met-Gly- | -Trp-Met-Asp-Phe-NH_2

PENTAGASTRIN (SYNTHETIC)

$\qquad\qquad\qquad\qquad\qquad$ C(CH$_3$)$_3$-OCO-NH-CH$_2$-CH$_2$-CO- | -Trp-Met-Asp-Phe-NH_2

SECRETIN

1 2 3 4 5 6 7 8 ·9 10 11 12 13
His-Ser-Asp-Gly-Thr-Phe-Thr-Ser-Glu-Leu-Ser-Arg-Leu-

14 15 16 17 18 19 20 21 22 23 24 25
Arg-Asp-Ser-Ala-Arg-Leu-Gln-Arg-Leu-Leu-Gln-Gly-

26 27
Leu-Val-NH_2

Note: The 14 amino acids in italics occupy the same position, counting from the N-terminus, as do those in glucagon. There is no active group, the whole molecule being required.

quantitative properties. The entire molecule is a very potent stimulant of acid secretion, but it is only a weak stimulant of gallbladder contraction.

Gastrin is synthesized and stored in cells, called G cells, in the antral mucosa of the stomach and in the duodenum and upper jejunum. Gastrin is released into the blood from the cells of the antral mucosa by the following stimuli:

1. During the cephalic phase of digestion, cholinergic impulses in vagal nerves release gastrin. Hypoglycemia, a blood sugar concentration of about 45 mg per 100 ml, acting through the hypothalamus, causes vagally stimulated release of gastrin. This fact forms the basis for a test of the completeness of vagotomy. If the stomach fails to

TABLE 5-2.   MAJOR PHYSIOLOGIC ACTIONS OF GASTRIN,
CHOLECYSTOKININ AND SECRETIN

GASTRIN

Increases resting pressure in lower esophageal sphincter
IMPORTANT: Stimulates acid secretion by the oxyntic cells
which in turn stimulates secretion of pepsinogen by chief
cells through local reflex
IMPORTANT: Increases gastric antral motility
Weakly stimulates enzyme and bicarbonate secretion by
pancreas, contraction of gallbladder
IMPORTANT: Has trophic effects on gastric mucosa

CHOLECYSTOKININ

Weakly stimulates gastric secretion of acid
IMPORTANT: Competitively inhibits gastrin-stimulated secretion of acid
IMPORTANT: Strongly stimulates enzyme secretion by pancreas
Weakly stimulates secretion of bicarbonate by pancreas, BUT
IMPORTANT: Strongly potentiates effect of secretin in stimulating
bicarbonate secretion by pancreas
IMPORTANT: Strongly stimulates contraction of gallbladder
Stimulates duodenal secretion and motility

SECRETIN

Stimulates pepsinogen secretion
IMPORTANT: Stimulates secretion of bicarbonate by pancreas and liver;
synergistic with CCK
IMPORTANT: Noncompetitively inhibits gastrin-stimulated secretion of acid
IMPORTANT: Inhibits gastric and duodenal motility
Inhibits lower esophageal sphincter
Has metabolic effects similar to those of glucagon

secrete acid when adequate hypoglycemia is induced by
insulin administration, vagotomy is presumed to be complete.
2. Distention of the gastric antrum causes release of gastrin.
The effect is mediated by a cholinergic reflex in the intrinsic plexuses.
3. Secretagogues in contact with the pyloric glandular mucosa cause the release of gastrin. The most potent are
amino acids and polypeptide digestion products of protein.
Aliphatic alcohols, of which ethanol is the most effective,
also cause gastrin release. Carbohydrates are ineffective.

Release of gastrin is greatest when the pyloric glandular mucosa is bathed by a neutral solution. As the pH of the solution becomes less, release of gastrin in response to any stimulus is progressively inhibited until, at about pH 1.5, inhibition of release is complete. Inhibition of gastrin release by acid is an important part of the control of acid secretion.

Factors governing release of gastrin from the intestinal mucosa are poorly understood.

Fat in contact with the duodenal mucosa causes the gallbladder to contract, and in the 1920's the effect was demonstrated to be hormonally mediated. The responsible agent was called *cholecystokinin*. Much later, secretion of enzymes by the pancreas in response to irrigation of the duodenum with protein digestion products was demonstrated to be mediated by a hormone, and the as yet unidentified hormone was called *pancreozymin*. The two actions were eventually shown to be different properties of the same molecule, now called cholecystokinin-pancreozymin, or, more conveniently, CCK-PZ.

CCK-PZ is a polypeptide chain of 33 amino acids, and the last four, counting from the N-terminus, are the same as those in the corresponding positions in gastrin. Therefore, CCK-PZ has the same qualitative spectrum of actions as gastrin. CCK-PZ's long chain of 29 amino acids attached to the active group confers on it specific quantitative actions. Thus gastrin is a strong stimulant of acid secretion but a weak stimulant of gallbladder contraction, but CCK-PZ is a weak stimulant of acid secretion and a strong stimulant of gallbladder contraction.

CCK-PZ is released into the circulation from cells in the duodenal mucosa. The most potent stimuli are fat and protein digestion products.

Secretin is a polypeptide containing 27 amino acids. It has no subunit with hormonal activity; the entire molecule is required. Fourteen of its amino acids occupy the same position, counting from the N-terminus, as do those of glucagon, and secretin has metabolic effects similar to those of glucagon.

Secretin is released into the circulation from cells in the duodenal mucosa by acid in contact with the mucosa. The amount of secretin released is proportional to the amount of acid flowing through the duodenum.

The action of secretin in stimulating secretion of bicarbonate-containing juice from the pancreas and liver is strongly potentiated by CCK-PZ.

During the course of embryologic development, cells that will be parents of the hormone-secreting cells of the viscera migrate from the neural crest to the organs of the digestive tract, and there they differentiate into cells with specific functions. Some become cells in the Isles of Langerhans which secrete insulin; others become those secreting glucagon. Still others become gastrin- and secretin-secreting cells. Errors in differentiation sometimes occur, and cells in the pancreas and other parts of the gut may synthesize and secrete gastrin continuously and at a very high rate. Consequently, the oxyntic glandular mucosa secretes acid continuously in large amounts. This acid often overwhelms the capacity of the duodenum to deal with the acid, and the pH of intestinal contents may be very low as far as the midjejunum. Duodenal and jejunal mucosa becomes severely ulcerated. Pancreatic lipase is denatured in the acid medium, and there is maldigestion of fat leading to steatorrhea. When the aberrant cells form discrete tumors, they may be surgically removed, and excessive acid secretion is alleviated. When they are diffusely distributed, complete gastrectomy is the only successful treatment.

Gastrin is a trophic hormone for the oxyntic cells, and in persons who have continuing high concentrations of gastrin in the plasma there is hypertrophy and hyperplasia of the oxyntic cells in the gastric mucosa. The capacity to secrete acid is increased.

Other discrete or diffuse tumors in the pancreas continuously produce either secretin or a secretinlike hormone. The patient's pancreas and liver pour enormous amounts of bicarbonate-containing fluid into the small intestine. The resulting state of profuse diarrhea is called "pancreatic cholera." In some instances, tumors have been found to se-crete both gastrin and the secretinlike hormone.⌐

6. GASTRIC SECRETION

The two main divisions of the gastric mucosa are the oxyntic glandular mucosa and the pyloric glandular mucosa (Fig. 6-1). The first, which secretes acid, covers the body of the stomach, and the second covers the antrum. The distribution of the two types of mucosa, however, does not exactly correspond to the two divisions of the stomach based on the nature of the muscular layer.

The surface of the oxyntic glandular area is covered with a layer of tall, columnar surface epithelial cells that contain and secrete mucus. The surface is thickly studded with pits into which long tubular glands empty. At the junction of pit and gland, are the neck chief cells. These, too, contain and secrete mucus, but their most important function is to serve as parent cells for the replacement of other cells of the gastric mucosa. The chief cells forming the walls of the tubules are the ones which synthesize and secrete the enzyme precursor, pepsinogen. Along the outside wall of the glands are the cells which secrete hydrochloric acid. Because they are on the wall, they are called parietal cells (Latin, *paries* = wall), and because they secrete acid they are called (oxyntic cells) (Greek, *oxy* = sharp). The latter name will be used here. In man the oxyntic cells also secrete the intrinsic factor, necessary for absorption of vitamin B_{12} in the terminal ileum. Other cells of the oxyntic glandular area contain large amounts of histamine, 5-hydroxytryptamine and heparin. Liberation of these compounds is important in pathologic conditions. The oxyntic glandular area also has a large histamine-forming capacity.

The surface of the pyloric glandular mucosa is also covered with surface epithelial cells that contain and secrete mucus. Secretion of mucus is copious during digestion of a meal, and

41

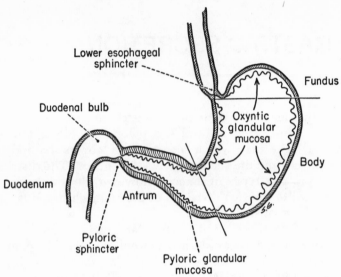

Fig. 6-1.—The parts of the stomach and duodenum. The border between the pyloric glandular mucosa and the oxyntic glandular mucosa does not exactly coincide with the border between antrum and body.

mucus appears to be an important lubricant of the pyloric glandular mucosa. Cells of the pyloric glands secrete a small amount of a neutral fluid roughly similar to an ultrafiltrate of plasma. They also secrete a small amount of pepsinogen, which is chemically different from that secreted by the oxyntic glandular mucosa.

The oxyntic glandular mucosa secretes a juice containing H^+, Cl^-, Na^+ and K^+. As the rate of secretion increases, the concentration of H^+ rises, and the concentration of Na^+ falls. At the highest rate of secretion, the fluid collected from the stomach is a nearly isotonic solution containing HCl at approximately 145 mN (millinormal) and KCl at approximately 10 mN (Fig. 6-2).

Samples of gastric juice taken from the stomach may have been diluted and partially neutralized by swallowed food and saliva or by regurgitated duodenal contents. When these

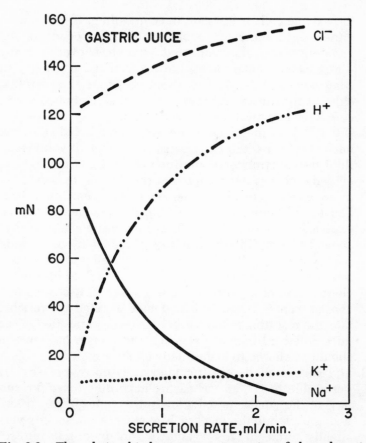

Fig. 6-2.—The relationship between concentration of electrolytes in the gastric juice of a normal young man and the rate of secretion. Secretion was stimulated by intravenous infusion of histamine at various constant rates, and the juice to be analyzed was collected after steady rate of secretion had been reached. (Adapted from Nordgren, B.: Acta Physiol. Scandinav., Suppl. 202, 1963.)

factors are eliminated, the composition of gastric juice varies with the rate of secretion. There are three explanations for this:

1. Most physiologists believe that the acid fluid secreted by

the oxyntic cells is constant in composition. If this is true, either of the other two explanations will account for the variability of the fluid collected. There has never been any direct experimental proof, however, of the constancy of composition of the fluid as it emerges from the secreting cells. Approximate constancy is probable, but absolute constancy remains to be demonstrated.

2. Other cells of the mucosa, the surface epithelial cells, the neck chief cells, the chief cells, secrete small volumes of fluid the electrolyte composition of which is thought, on evidence that is not completely convincing, to be similar to an ultrafiltrate of plasma. As such, it contains bicarbonate at about 24 mN, and bicarbonate neutralizes an equivalent amount of acid. The fluid also contains Na^+ at about 145 mN. This fluid is supposed, again on not completely convincing evidence, to be secreted at a constant rate when acid secretion is stimulated. If a fluid having the composition of an ultrafiltrate of plasma and secreted at a constant rate is mixed with acid juice secreted at a variable rate, the resulting relationship between composition of the juice collected from the stomach and its rate of secretion should be similar to that actually found (Fig. 6-2).

3. The surface of the gastric mucosa is only very slightly permeable to H^+, and as gastric juice flows over the surface of the stomach, acid slowly leaves the juice by diffusing back into the mucosa. The mucosa is also only very slightly permeable to Na^+ and Cl^- contained in its interstitial fluid. As acid gastric juice flows over the surface of the stomach, Na^+ slowly diffuses into gastric juice in exchange for H^+, and an additional small amount of Na^+ and Cl^- diffuse together from interstitial fluid into gastric juice. These processes of diffusion may also explain the relationship between the rate of secretion of gastric juice and its composition.

Explanations 2 and 3 are not mutually exclusive.

[Gastric mucosal cells contain an electrogenic chloride pump. When there is little or no stimulus for secretion, this pump is minimally active, but because it continues to secrete Cl^- into the lumen at a low rate, the luminal surface of the oxyntic glandular mucosa is negative with respect to the serosal surface or the blood by about 40 to 60 mV.

When the oxyntic cells are stimulated to secrete acid, the electrogenic H^+ pump becomes active, and the Cl^- pump increases its rate of pumping so that H^+ and Cl^- are secreted together. Then the potential difference across the mucosa falls slightly. (In experimental conditions in vitro, the gastric mucosa can be made to secrete H^+ without any accompanying Cl^-, and in that case the luminal surface of the mucosa becomes positive with respect to the serosal surface.)

The chemical mechanism by which oxyntic cells secrete acid is not known, but whatever the mechanism may be, neutrality of the interior of the cells is maintained by replacing the H^+ secreted with H^+ derived from carbonic acid (Fig. 6-3). Carbon dioxide from the blood or from metabolism is hydrated to carbonic acid, and the hydration is catalyzed by carbonic anhydrase. Carbonic acid ionizes, giving an H^+ which replaces the one secreted and an HCO_3^- which passes into the plasma. The HCO_3^- replaces the Cl^- taken from plasma to accompany H^+ in the secreted juice. The means by which K^+ enters the juice is unknown. Water flows from the plasma into the juice as the ions are secreted, and the juice is very nearly isotonic with the plasma in gastric capillaries.

In the process of secretion of acid juice, the metabolic machinery of the oxyntic cells manufactures new ions, H^+ and HCO_3^-, which are osmotically active. Consequently, as secretion proceeds, the total osmotic pressure of the plasma rises.

When gastric juice is lost, either by vomiting or by drainage, metabolic alkalosis results, because for each H^+ lost in the juice, one HCO_3^- has been retained in the blood. Loss of gastric juice also causes dehydration and shrinkage of extracellular fluid volume, and loss of K^+ may result in a negative potassium balance with all its consequences.

Three classes of exogenous compounds stimulate gastric secretion:

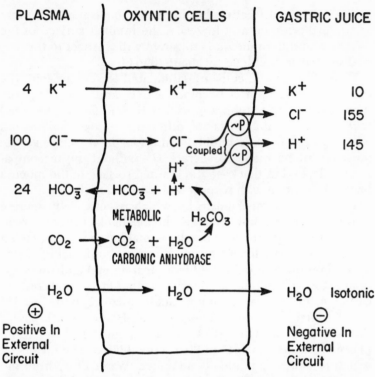

Fig. 6-3.—Processes involved in secretion of acid by an oxyntic cell. When the cell is not stimulated, the electrogenic chloride pump is minimally active, and a potential difference is established across the cell. When the cell is stimulated, the electrogenic hydrogen ion pump secretes H^+, and the potential difference falls slightly. The chloride pump, which is coupled with the hydrogen ion pump, increases its activity, and the two ions are secreted together. Flow of water through the cell makes the secreted juice nearly isotonic. Carbonic acid derived from metabolic and plasma carbon dioxide furnishes hydrogen ions to replace those secreted, and bicarbonate to replace chloride ions removed from plasma. The numbers in the margins give the approximate concentrations of the ions in plasma and gastric juice in milliequivalents per liter.

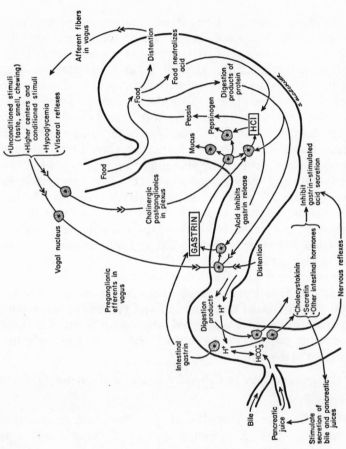

Fig. 6-4.—The major factors controlling gastric secretion.

1. *Cholinergic drugs.* Long-lived analogs of acetylcholine such as bethanechol, when administered by injection, stimulate secretion of HCl, pepsinogen and mucus.

2. *Histaminergic drugs.* Histamine and its analogs, when administered by injection, strongly stimulate the secretion of HCl and weakly stimulate the secretion of pepsinogen. In the "augmented histamine test," the side-effects of histamine are minimized by administration of an antihistaminic drug, and a dose of histamine thought to stimulate maximally is given. The secretory response is judged to be a measure of the stomach to secrete acid. The gastric mucosa contains large stores of histamine and has a high histamine-forming capacity. When the gastric mucosa is injured, histamine is liberated, and it stimulates acid secretion. Whether histamine has any role in normal control of acid secretion is undecided.

3. *Gastrinlike drugs.* Gastrin and its analogs, when administered by injection, stimulate acid secretion. Because HCl on the surface of the mucosa itself stimulates secretion of pepsinogen, juice collected after gastrin administration contains pepsin. Pentagastrin is the most commonly used analog of gastrin.

Gastric secretion is naturally stimulated in the following ways (Fig. 6-4):

1. *Cephalic phase.* Stimuli acting in the head, among them smelling, tasting, chewing and swallowing food, cause impulses to flow along the vagus nerves to the stomach. In the body of the stomach, these impulses cause the release of acetylcholine near secretory cells, thereby stimulating secretion of HCl, pepsinogen and mucus. In the antrum, the impulses promote the liberation of gastrin, provided the pyloric glandular mucosa is not bathed by an acid solution.

2. *Gastric phase.* Food in the stomach stimulates gastric secretion by three means:
   (a) Distention of the stomach stimulates mechanoreceptors the afferent fibers of which travel centrally in the vagus

nerve. Resulting efferent discharges, also in the vagus nerve, stimulate gastric secretion.

(b) Distention of the stomach stimulates mechanoreceptors the afferent impulses of which increase activity in the intrinsic plexuses of the body of the stomach and stimulate gastric secretion by a local reflex. Components of the food, particularly polypeptide digestion products of proteins, stimulate chemoreceptors, which also stimulate copious secretion of acid through a local reflex.

(c) Distention of the antrum of the stomach stimulates release of gastrin from the pyloric glandular mucosa, provided the mucosa is not bathed by an acid solution. Secretagogues, particularly ethanol and digestion products of protein, also stimulate gastrin release if they are in neutral solution.

3. *Intestinal phase.* Chyme in the intestine stimulates secretion of acid juice by the stomach. At least one mediator is gastrin released from the duodenal and upper jejunal mucosa, and there are probably others.

Gastric secretion is inhibited by the following means:

1. *Exogenous drugs.* Because acetylcholine stimulates acid and pepsinogen secretion, atropine and atropinelike drugs inhibit it. The common run of antihistaminic drugs that are useful in controlling many of the effects of histamine have little or no effect on gastric secretion. At least one newer antihistaminic drug, however, does inhibit acid secretion. Inhibitors of carbonic anhydrase, when given in massive amounts, inhibit acid secretion, but this is unimportant.

2. *Reflexes.* Acid, fat digestion products and solutions of high osmotic pressure in the duodenum and upper jejunum inhibit gastric secretion, in part through nervous reflexes that are poorly understood. There are noncholinergic, nonadrenergic fibers in the vagus nerve which inhibit gastric secretion when stimulated, and sympathetic nerves and intrinsic plexuses may also participate.

3. *Hormones.* CCK-PZ released from the intestinal mucosa competitively inhibits gastrin-stimulated secretion of acid.

The basis for this is competition for receptor sites on the secreting cells. In order to stimulate copious secretion of acid, gastrin must combine with a receptor on the cell. CCK-PZ has the same amino acid sequence in its active group, and it too combines with the same receptor. CCK-PZ, however, is only a weak stimulant of acid secretion. By occupying the receptor site, CCK-PZ denies the site to gastrin, a much stronger stimulant of acid secretion. Secretin from the intestinal mucosa also inhibits gastrin-stimulated acid secretion, but because its structure has no relation to that of gastrin, secretin inhibition is noncompetitive.

Before the isolation and identification of CCK-PZ and secretin, crude extracts of the duodenal mucosa were found to inhibit gastric secretion and motility. The supposed active principle was called *enterogastrone*. At least part of the effects of enterogastrone can be attributed to the extracts' content of CCK-PZ and secretin, but there are other compounds in such extracts that also inhibit gastric secretion and motility. Their physiologic role is at present unknown.

The rate of acid secretion rises to a maximum in the second half hour after the beginning of a meal, and it declines slowly over the next several hours. The total amount of acid secreted in response to a meal is directly proportional to the protein content of the meal (Fig. 6-5). The reason is that protein is the best buffer contained in food. Dilution and neutralization of residual acid in gastric contents at the beginning of a meal allow gastrin to be released. Gastrin stimulates secretion of acid, and acid titrates the buffers of the food. Eventually, when the buffers have been titrated to a low pH, acid in contact with the pyloric glandular mucosa inhibits further release of gastrin, and a major stimulant of acid secretion is withdrawn. Any other buffer or neutralizing agent added to food has the same effect as protein in governing the amount of acid secreted.

When the stomach has emptied and stimuli for acid secretion have been withdrawn, the stomach is left with a small

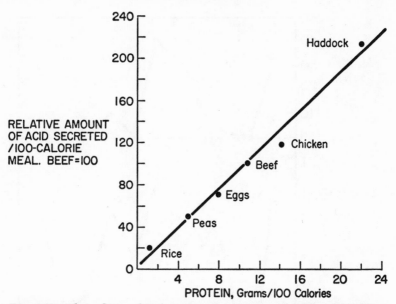

Fig. 6-5.—The relationship between the amount of acid secreted by the stomach during digestion of a meal and the protein content of the food. The amount of acid secreted in response to a 100-calorie meal of beef is put at 100 and is plotted against the protein content of the beef used, 10.7 gm per 100 calories. The relative amounts of acid secreted in response to 100-calorie meals of the foods named are plotted against the protein content of the foods. Points obtained with 24 other foods having a wide range of protein content fall close to the line drawn in the figure. (Adapted from Saint-Hilaire, S., *et al.*: Gastroenterology 39:1, 1960.)

volume of acid juice. The relation between acidity of gastric contents and meals is shown in Figure 6-6.

In some persons, the oxyntic glandular mucosa atrophies, losing its ability to secrete acid and pepsinogen. Because there is no acid in the stomach to inhibit the release of gastrin, the plasma concentration of gastrin is very high, but it falls abruptly when some 100 mN HCl is drunk. Acid and pepsin are not essential for protein digestion. Absorption of iron is better if the stomach is capable of secreting acid, and persons with gastric mucosal atrophy may have some degree

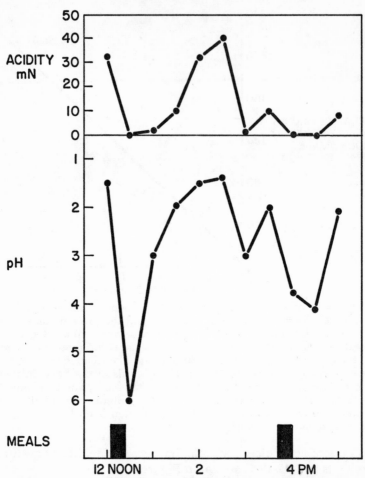

Fig. 6-6.—The acidity of contents removed from the human stomach by tube expressed as millinormal acid and as pH. At 12 noon, there was only a small amount of acid fluid in the stomach. After a luncheon of fish, potato, carrot, fruit, custard and tea, the acidity of the contents of the stomach was very low. The acidity gradually climbed as acid was secreted in response to the meal and as the stomach emptied. Acidity fell once more at 4 P.M., when bread and butter were eaten with tea. (Adapted from James, A. H., and Pickering, G. W.: Clin. Sci. 8:181, 1949.)

of iron deficiency. The most important consequence of mucosal atrophy, however, results from its failure to secrete intrinsic factor necessary for absorption of vitamin B_{12} in the terminal ileum, and persons with deficiency of intrinsic factor have pernicious anemia. Many persons with gastric atrophy have antibodies to oxyntic cells in their blood, and gastric atrophy may be the result of an autoimmune process.

All the cells of the gastric mucosa turn over rapidly, and the whole mucosa is replaced in about 3 days. Approximately a half million cells desquamate into the lumen of the stomach each minute, and they can be identified in gastric washings. The neck chief cells are the parents of all new cells. After they divide, the daughter cells migrate toward the surface of the mucosa and differentiate into surface epithelial cells. Other daughter cells migrate down the glands, differentiating into chief and oxyntic cells. The ability of the mucosa to renew itself is particularly important when the stomach is injured.

The permeability characteristics of the gastric mucosa are different from those of the rest of the gut. For instance, it is almost entirely unaffected by osmotic gradients. Pure water in the stomach is not absorbed, and if the contents of the stomach are hypertonic, little water moves from blood into the lumen. This is in contrast with the duodenum where hypotonic or hypertonic solutions are quickly brought to isotonicity.

The gastric mucosa is normally only very slightly permeable to the acid which it secretes. This property, called the *gastric mucosal barrier*, accounts for the stomach's ability to contain acid without injuring itself. Many substances, however, break the gastric mucosal barrier. The most important of these are salicylates: aspirin and salicylic acid in acid but not in neutral solution. Other important compounds which break the barrier are ethanol and regurgitated bile acids and lysolecithin of duodenal contents. In some otherwise normal persons, the barrier may be, for unknown reasons, weak, and it frequently breaks in patients with severe injuries or infections. When the barrier is broken, acid can diffuse back into the mucosa with serious pathophysiologic consequences (Fig. 6-7).

1. Acid diffusing back into the mucosa stimulates motility of the stomach by acting on the intrinsic plexuses. Strong contractions by an indurated stomach may give rise to pain.
2. Acid diffusing back into the mucosa stimulates secretion of pepsinogen.
3. Acid diffusing back into the mucosa releases histamine from mucosal stores and increases the mucosa's histamine-forming capacity. Histamine released stimulates acid secretion with the result that a mucosa into which acid is diffusing simultaneously secretes acid.

Fig. 6-7.—Pathophysiologic consequences of the back-diffusion of acid through the broken gastric mucosal barrier.

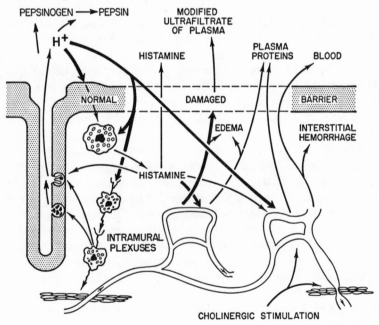

4. Histamine, and perhaps other substances liberated during mucosal injury, cause increased capillary permeability and vasodilation. Protein-containing fluid pours from the capillaries into the interstitial spaces, and the mucosa becomes edematous. Large volumes of interstitial fluid exude from the mucosa. The process may end in capillary stasis.
5. A large quantity of plasma proteins may be shed by the injured mucosa.
6. Bleeding ranging from minor superficial hemorrhage to exsanguination occurs.

7. GASTRIC MOTILITY

With each swallow as a meal is eaten, there is a slight receptive relaxation of the body of the stomach so that the stomach accommodates itself to the meal with little rise in intragastric pressure. Receptive relaxation is a vagally mediated reflex. Throughout the digestion of a meal, the contents of the body of the stomach remain undisturbed, for only feeble mixing movements, if any, occur in its wall. Peristaltic waves move over the antrum to the pyloric sphincter, mixing food with digestive juices and slowly propelling the mixture into the duodenum. As the volume of the contents of the body of the stomach decreases, the wall of the body gently contracts.

Because the contents of the body are not mixed, digestion of protein by pepsin is slight or absent. The acid juice secreted by the oxyntic glandular mucosa flows around the mass of undisturbed food to the antrum, where it is thoroughly mixed with small amounts of food by strong peristaltic contractions of antral muscle. Therefore, digestion of starch by salivary ptyalin continues in the body of the stomach but is stopped by acid in the antrum, whereas digestion of protein begins in the antrum.

As the stomach is filled by a meal, the concentration of gastrin rises in the plasma. Circulating gastrin increases the resting pressure of the lower esophageal sphincter. Thus, at the time the stomach is distended and acid is being secreted, tighter closure of the sphincter reduces the possibility that acid gastric contents will be regurgitated into the esophagus.

Peristalsis is governed by a wave of partial depolarization which begins in a group of pacemaker cells in the longitudinal muscle layer high on the greater curvature of the stomach. The wave sweeps over the longitudinal layer toward the pylorus, and it almost completely dies out at the pyloric sphincter. This BER, as it moves over the longitudinal muscle, may or may not be accompanied by contraction of the underlying circular muscle. If the circular muscle contracts, a peristaltic wave moves over the stomach in step with the BER; if the circular muscle does not contract, the BER sweeps on anyway with no visible sign of its presence. The frequency with which the BER originates is close to 3 per minute, so the rhythm of peristalsis in the stomach is the same. The velocity of the BER over body and antrum is so slow, about 1 cm per second, that two or three successive peristaltic waves may be seen at one time. The wave accelerates as it reaches the terminal antral segment, with the result that the terminal antrum and the pyloric sphincter contract almost simultaneously.

If contraction of the circular muscle does accompany the BER, the contraction may be weak or strong. The BER is a wave of partial depolarization, and consequently it is a current sink. Current flows from the surface of the circular muscle fibers to the sink, and this electrotonic flow of current tends to depolarize the circular muscle. Whether or not the circular muscle fibers are depolarized sufficiently to bring them to threshold so that they have action potentials and increase in tension depends upon the excitability of the circular muscle fibers at the time. If there is a lot of vagal excitatory activity, threshold will be low. Many circular muscle fibers will reach threshold, spike and contract. The wave of peristalsis will be strong. If there is little vagal activity or if there is sympathetic inhibitory activity, threshold will be high. Few or no circular muscle fibers will reach threshold, and peristaltic contractions will be weak or absent.

The body of the stomach is a hopper, and its contents are gradually fed into the antrum (Fig. 7-1). No matter how strong an antral peristaltic wave may be, its ring of contraction is never closed. As a wave advances, it mixes food and digestive juices within the antrum, and the mixture escapes backward through the open ring, only to be tumbled again by the next peristaltic wave.

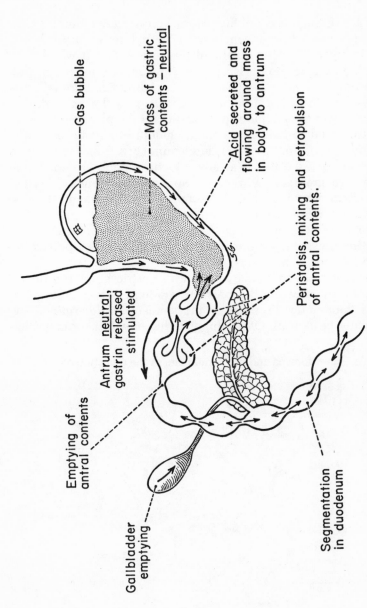

Gas bubble

Mass of gastric contents – <u>neutral</u>

<u>Acid</u> secreted and flowing around mass in body to antrum

Peristalsis, mixing and retropulsion of antral contents.

Antrum <u>neutral</u>, gastrin released stimulated

Emptying of antral contents

Segmentation in duodenum

Gallbladder emptying

Fig. 7-1.—The stomach and duodenum at the beginning of the digestion of a meal.

At rest, the muscle of the pyloric sphincter is either relaxed or only very slightly contracted. The pyloric canal is closed and empty. There is a zone of pressure within it which is a bit higher than the pressure in the antrum or duodenum, and therefore no chyme flows in either direction through the canal. As the peristaltic wave advances in the antrum, the viscosity of the chyme it propels causes the pressure in front of the canal to rise, and a small amount of chyme may pass through the canal into the duodenum (Fig. 7-2). How much chyme passes through at any one time depends upon the pressure gradient between antrum and duodenum. As the peristaltic wave reaches the terminal antral segment, the pyloric sphincter contracts and abruptly cuts off passage of chyme into the duodenum. The sphincter then relaxes to its resting pressure and remains empty until the next peristaltic wave comes along.

Factors that influence the rate of gastric emptying do so chiefly by affecting the pressure gradient between stomach and duodenum. Pressure in the stomach is determined by tension in its wall. Other things being equal, the greater the

Fig. 7-2.—Stomach and duodenum late in the digestion of a meal.

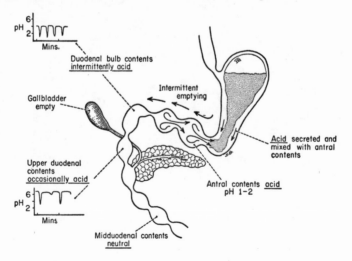

volume of gastric contents, the greater is the tension. Consequently, for a liquid meal, the rate of emptying is greater the larger the volume of gastric contents, and the rate of emptying is greater at the beginning of the process than at the end (Fig. 7-3). For this reason, the concept of "emptying time," the time the last trace of a meal leaves the stomach, has little meaning.

The rate of emptying is governed by the ability of the duodenum to deal with the chyme delivered to it. The three

Fig. 7-3.—The volume of gastric contents of a normal man plotted against time, showing that the rate of emptying is a linear function of the square root of the volume remaining in the stomach. The reason for the square root relation may be that the circumferential tension in the wall of a cylinder is proportional to the square root of its volume. The osmotic pressure of the hypertonic and hypotonic meals was adjusted by adding sucrose to a mixture of citrus pectin and water. The complete liquid meal was a milk product to which sugar had been added. (Adapted from Hunt, J. N., and Spurrell, W. R.: J. Physiol. 113:157, 1951; Hunt, J. N.: Gastroenterology 45:149, 1963; and Hopkins, A.: J. Physiol. 182:144, 1966.)

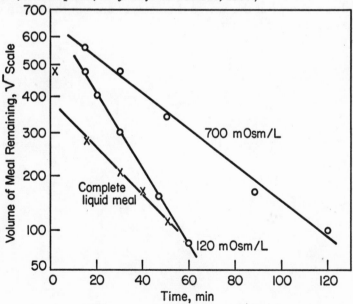

most important qualities of the chyme are its acidity, its osmotic pressure and its fat content. Gastric emptying is slowed until chyme in the duodenum is neutralized and made isotonic and until its fat content is reduced. Slowing of gastric emptying is acccomplished by reflexes and hormones originating in the duodenum; these inhibit gastric motility, relaxing the wall of the stomach and diminishing the strength of peristaltic contractions.

Solids or semisolids are emptied much more slowly than liquids. A solid piece of food is denied entry into the pyloric canal; it is pushed backward and forward in the antrum until it is sufficiently eroded to be able to pass through the canal.

When the stomach is finally empty after a meal, peristaltic contractions die out, and the stomach is quiet for a while. As hunger before the next meal advances, however, peristaltic waves again sweep over the nearly empty antrum. The stimulus is increased vagal activity aroused by a fall in glucose utilization in the hypothalamus. A person may experience hunger pangs during a particularly vigorous series of peristaltic waves, but the waves themselves are not the origin of the sensation of hunger.

8. PANCREATIC SECRETION

The external secretion of the pancreas contains two components:

1. One is an aqueous secretion of highly variable volume whose most important characteristic is that it contains a high concentration of bicarbonate. The liver secretes a similar fluid, in addition to the fluid secreted with bile acids, and secretion of this fluid by the liver is under the same control as that of the aqueous secretion of the pancreas. When the aqueous secretion of the pancreas is discussed, it is to be understood that the aqueous, bile acid-independent secretion of the liver is included.

2. The other is a solution of small volume containing all the enzymes and enzyme precursors synthesized and secreted by the acinar cells of the pancreas. This secretion is swept into the duodenum by concurrently secreted aqueous juice.

The aqueous component, secreted by the cells of the intercalated ducts and modified as it passes down the pancreatic ducts, is an isotonic solution the major cation of which is sodium. It also contains a small amount of potassium, calcium and magnesium. The anions are bicarbonate and chloride the sum of which equals that of the cations. As the rate of secretion of human pancreatic juice increases, the rate of output of bicarbonate rises, and the rate of output of chloride correspondingly falls (Fig. 8-1). The function of this secretion is to neutralize acid from the stomach.

Minor stimuli of pancreatic aqueous secretion are impulses arriving by vagal fibers and gastrin liberated from the gastric antral mucosa. Consequently, there is a small cephalic phase of pancreatic and gastric secretion. Major stimuli are the hormones secretin and CCK-PZ.

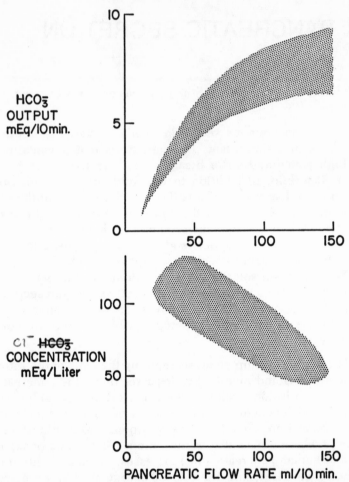

Fig. 8-1.—Top, the relationship between the bicarbonate output of the pancreas of normal human subjects and the rate of secretion of pancreatic juice collected by duodenal aspiration when pancreatic secretion was stimulated by intravenous administration of secretin. Most observed points fall within the stippled area. **Bottom,** the relationship between the bicarbonate concentration of pancreatic juice collected from normal human subjects by duodenal aspiration and the rate of secretion. Most observed points fall within the stippled area. (Adapted from Wormsley, K. G.: Gastroenterology 54:197, 1968.)

The hormone secretin is liberated from the duodenal mucosa by acid, and the amount of secretin liberated is proportional to the amount of acid entering the duodenum. Secretin is a potent stimulant of pancreatic aqueous secretion, and in large doses it causes maximal secretion of the aqueous component. In the normal course of digestion, however, not enough secretin is liberated to account for the amount of pancreatic juice secreted. This is because most of the acid arriving in the duodenum is promptly neutralized, and only a small segment of duodenal mucosa is exposed to acid.

The hormone CCK-PZ is in itself a weak stimulant of pancreatic aqueous secretion, but it exerts a powerful synergistic effect with secretin. CCK-PZ is liberated when fat and protein digestion products come into contact with the duodenal mucosa, and this occurs at the same time acid from the stomach enters the duodenum. Consequently, the two hormones, secretin and CCK-PZ, are liberated at the same time, and acting together, they stimulate secretion of the aqueous component of pancreatic juice.

The major enzymes synthesized and secreted by the acinar cells of the pancreas are the following:

1. *Proteolytic enzymes.* These include trypsinogen, chymotrypsinogen, procarboxypeptidase and proaminopeptidase. They are all inactive as they are secreted. An enzyme secreted by the duodenal mucosa, enterokinase, acts on trypsinogen to convert it to the actively proteolytic enzyme trypsin. Trypsin in turn autocatalytically activates trypsinogen and converts chymotrypsinogen and the propeptidases into active enzymes. Pancreatic juice contains a low concentration of trypsin inhibitor.

2. *Amylolytic enzyme.* This is an amylase which catalyzes the hydrolysis of raw or cooked starch and glycogen. The enzyme breaks the $\alpha-1,4$ glucosidic bonds but not the $\alpha-1,6$ bond. The result is that the products of starch hydrolysis are glucose, short straight-chain oligosaccharides (maltose and the like) and isomaltose containing a branching point.

3. *Lipolytic enzyme.* This is a lipase which catalyzes the hy-

drolysis of the 1 and 1' ester bonds of triglycerides, especially those containing long-chain fatty acids.

There are many other enzymes including elastase and ribonuclease in pancreatic juice.

All the enzymes are synthesized by the rough endoplasmic reticulum of the acinar cells, and they are formed into granules stored at the apex of the cells. Each granule apparently contains all the enzymes. Although there is some variability in the proportions in which the enzymes are secreted, all enyzmes are secreted together roughly in parallel. There is no evidence that in man the enzyme composition of pancreatic juice changes radically in adaptation to changes in diet.

9. FUNCTIONS OF THE SMALL INTESTINE

The duodenum responds to the chyme delivered to it from the stomach by the following processes:

(1) Acid in the chyme is neutralized, and chyme is neutral as it leaves the duodenum. (Chyme becomes more alkaline in the ileum as bicarbonate is secreted by the ileal mucosa.)

(a) The rate of emptying of the stomach is regulated so that no more acid is emptied into the duodenum than can be neutralized in the duodenal bulb and the first few centimeters of the duodenum.

(b) Secretin and CCK-PZ are released into the blood from the duodenal mucosa, and together they stimulate the pancreas and the liver to secrete bicarbonate-containing fluids.

(c) Some acid is absorbed through the duodenal mucosa or neutralized by bicarbonate in duodenal secretions. In the absence of pancreatic juice, the pH of chyme in the duodenum is about one pH unit lower than normal, which means the chyme is 10 times more acid.

(2) The osmotic pressure of duodenal contents is adjusted to isotonicity, and the chyme remains isotonic throughout the rest of the small intestine. Chyme emptied from the stomach may be hypotonic or hypertonic, depending on what is eaten. During digestion of food in the duodenum, a hypertonic solution may be produced. For example, starch has a negligible osmotic pressure, but when hydrolysis is catalyzed by pancreatic amylase many small molecules are rapidly liberated, and these raise the osmotic pressure of duodenal contents.

(a) Gastric emptying is slowed until duodenal contents become isotonic.

(b) Isotonicity is achieved by rapid flow of water from blood into duodenal contents.

3. The major digestion of food begins in the duodenum.

(a) CCK-PZ is released, and it stimulates the pancreas to secrete its enzymes. It also stimulates the gallbladder to contract and to empty concentrated bile into the duodenum. The result is that the enzyme and bile acid content of duodenal chyme rises abruptly about 20 minutes after the beginning of a meal.

(b) The contents of the duodenum are thoroughly mixed by segmental movements.

4. Absorption of digestion products begins in the duodenum. Glucose, galactose and other monosaccharides, amino acids and small polypeptides, 2-monoglycerides and free fatty acids are absorbed. As osmotically active particles are absorbed from duodenal contents, a corresponding amount of water is absorbed, and duodenal contents remain isotonic.

Gastric emptying is delayed by CCK-PZ and secretin at the same time these hormones inhibit acid secretion. Acidity and high osmotic pressure of duodenal contents stimulate duodenal motility (Fig. 9-1). Vigorous segmental movements spread the acidic or hypertonic chyme over a large area of duodenal mucosa, thereby facilitating rapid neutralization and adjustment of osmotic pressure. On the other hand, fat and fat digestion products depress duodenal motility as well as gastric emptying.

Surgical operations on the stomach and duodenum may destroy the duodenum's ability to regulate gastric emptying. For example, a gastrojejunostomy allows contents of the stomach to enter the upper small intestine rapidly. When gastric contents are mixed with pancreatic enzymes, starch and proteins are quickly hydrolyzed, and the contents of the small intestine become hypertonic. Water pours from the blood into the intestinal lumen, the intestine is distended and plasma volume falls. Because the normal means by which gastric emptying is delayed no longer operate, intes-

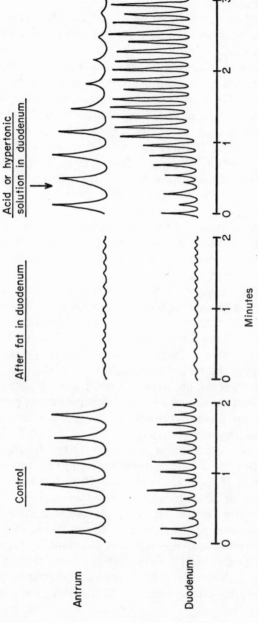

Fig. 9-1.—Motility of the human gastric antrum and duodenum influenced by the nature of duodenal contents. The control records show motility during the digestion of a mixed meal. Increases in pressure in the gastric antrum are caused by peristaltic waves the frequency of which is three times per minute, and irregular increases in pressure in the duodenum are caused by segmental movements occurring at a higher frequency. Soon after fat is placed in the duodenum, motility in both antrum and duodenum decreases. (Pressure variations in the record made after fat reflect respiratory movements.) Immediately after acid or a hypertonic solution is placed in the duodenum, antral motility decreases, and the strength of duodenal contractions greatly increases. (Adapted from records made by C. F. Code.)

tinal distention and fall in plasma volume continue unabated. These events, inelegantly termed *dumping*, may be followed by sweating, dizziness and fainting, all the results of a rapid fall in plasma volume.]

As chyme enters the duodenum, the motility of the small intestine is stimulated. [In contrast with the stomach, the chief movement of the small intestine is segmentation, not peristalsis. In segmentation, the circular muscle of the small intestine contracts in a series of rings, each several centimeters from the other (Fig. 9-2). The rings of contraction do not move along the intestine. Chyme is forced from the point of contraction in both directions into the intervening uncontracted segments. Shortly thereafter, the contracted ring of muscle relaxes, and the intervening, hitherto uncontracted segment contracts. Chyme is once more forced in both directions. The process of alternating contraction and relaxation continues indefinitely, with chyme being thrown backward and forward like a flying shuttle.

Segmentation is highly efficient in mixing chyme with digestive juices and in exposing all the chyme to the absorptive surface of the intestinal mucosa. Segmentation also moves chyme slowly down the small intestine, because there is a gradient of frequency of segmentation. The frequency is higher in the upper part of the small intestine than in the lower part. Consequently, chyme is moved from the upper to the lower part, because the chance that any particular part of the chyme will be pushed downward is greater than the chance that it will be pushed upward.

Frequency of segmentation is governed by the frequency of the BER in the local longitudinal muscle. The intrinsic frequency of the BER falls along the length of the small intestine. Response of the circular muscle, as expressed in segmental contraction, is enhanced by presence of chyme in the lumen. Distention of the intestine tends to cause it to contract, and distention of the inactive segment by chyme squirted into it facilitates its subsequent contraction. Segmentation occurs in vagotomized or sympathectomized persons, but it is influenced by extrinsic nerves. Vagal stimula-

Dog's jejunum segmenting

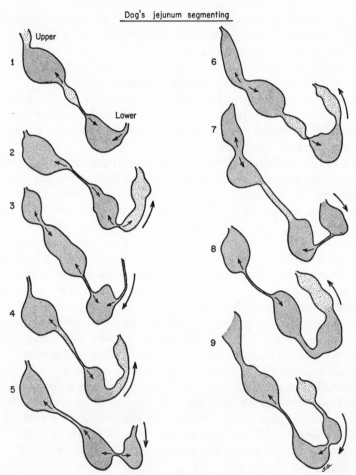

Fig. 9-2.—The process of intestinal segmentation. A dog had been fed a meal mixed with the x-ray contrast medium, barium sulfate, and continuous cinefluorographic pictures were taken as the contrast medium entered the upper jejunum. The tracings in this figure were made from the film to show the locus of the medium at approximately 1-second intervals. The *arrows* show the direction in which the jejunal contents were moving at the instant represented by the tracing. Progress of the contents into two distal segments over a period of about 9 seconds shows that intestinal contents can be moved from above downward by segmentation without peristalsis. (Adapted from a film made by H. C. Carlson.)

tion enhances segmentation, and cholinergic drugs are used to promote intestinal movement. Adrenergic and sympathetic influences inhibit segmentation.

Peristalsis also occurs in the human small intestine, but peristaltic rushes moving a long distance are definitely abnormal in man. Peristaltic waves occurring during the digestion of a meal may be no more than a train of two or more peristaltic contractions slowly moving approximately 10 centimeters before dying out.

In the interdigestive period, the jejunum is empty (as its name implies), and the ileum contains a slurry of undigested fibers and unabsorbed solutes. The terminal ileum is separated from the colon by the ileocecal sphincter. Pressure on the mucosa of the cecum by cecal contents causes the sphincter to contract, and distention of the terminal ileum causes it to relax. When a meal is being emptied into the duodenum, the duodenum begins to segment, and at the same time the ileum begins to segment as well. This concurrence of gastric and ileal activity is called the *gastroileal reflex*. As the chyme is pushed into the terminal ileum by segmentation, distention of the ileum causes the ileocecal sphincter to relax, and with each terminal ileal segmentation, a small squirt of chyme passes through the sphincter into the cecum. The cecum itself is aroused to activity, and the subsequent movements of the colon are said to be caused by the *gastrocolic reflex*. While food is in the stomach, gastrin is being released. Circulating gastrin decreases the strength of contraction of the ileocecal sphincter, allowing chyme to pass through it more easily.

In addition to being inhibited by distant events that cause widespread sympathetic adrenergic activity, the intestine is also inhibited by local injury. Pathologic dilatation of a part of the small intestine inhibits motility of the part of the intestine above it. This influence of one part of the intestine upon another is called the *intestinointestinal reflex*.

10. ABSORPTION OF WATER AND ELECTROLYTES

In addition to the water contained in food and drink, a large volume of water is added to gastrointestinal contents as salivary, gastric, pancreatic and biliary secretions each day. The total is probably 5 to 10 liters. Of this, all but about 500 ml is absorbed in the small intestine, and all but about 100 ml of the water entering the colon is absorbed.

When a steak meal having an initial volume of 645 ml is eaten, a total volume of 1,500 ml reaches the midjejunum. At the end of the jejunum, 750 ml is left; and in the ileum, the volume is reduced to 250 ml. Net volumes of fluid, however, have little meaning when intestinal handling of fluid is being considered; net volume movement across the intestinal mucosa is the resultant of very large movements of fluid and electrolytes in opposite directions across the intestinal mucosa (Fig. 10-1).

In contrast to what occurs in the stomach, pure water is very rapidly absorbed from the lumen of the small intestine. If the water is labeled with deuterium or tritium, however, water molecules are found to be moving in both directions during the process of absorption. There is a large movement from lumen to blood and a smaller one from blood to lumen, with net movement being absorption from the lumen. When water is in the lumen, Na^+ and associated anions, Cl^- and HCO_3^-, move from blood into the water in the lumen. As the water is absorbed, so is Na^+ and the anions.

When solutions containing NaCl are placed in the lumen of the intestine, unidirectional flow of water from blood to lumen is almost constant and independent of the osmotic pressure of the fluid in the lumen. Likewise, the unidirectional movement of Na^+ from blood to lumen is constant and

INTESTINAL LUMEN   BLOOD

Pure Water In Lumen

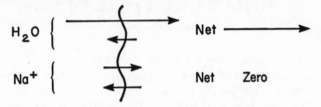

NaCl Up To 210 mN In Lumen

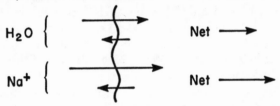

NaCl Above 210 mN In Lumen

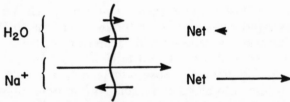

After NaCl Concentration Is Reduced Below 210mN,
Net Absorption Of Both
Na⁺ And H₂O Occur

Fig. 10-1.—The absorption of water and sodium by the small intestine as the resultant of opposed unidirectional fluxes.

independent of the concentration of Na^+ in the luminal fluid. The unidirectional flux of water from lumen to blood, however, is strongly influenced by the osmotic pressure of the solution. When the solution is hypotonic, unidirectional flux of water from lumen to blood is greater than the flux in the other direction, and net absorption occurs. As the solution is made more and more hypertonic, the unidirectional flux of water from lumen to blood decreases until, at a solute concentration approximately equal to a 210 mN solution of NaCl, the flux from lumen to blood is equal to that in the other direction, and no net absorption occurs. Above this solute concentration, the flux of water from lumen to blood is smaller than that in the opposite direction, and the volume of fluid in the lumen increases.

As the concentration of Na^+ in the lumen increases, the unidirectional flux of Na^+ from lumen to blood increases while that from blood to lumen remains constant. Therefore, net absorption of Na^+ increases with increasing luminal concentration (Fig. 10-2).

If the concentration of NaCl in the lumen is greater than about 210 mN, two processes occur at once: net flow of water is from blood to lumen, and net flux of Na^+ is from lumen to blood. Both of these reduce the concentration of NaCl in the lumen, and eventually its concentration falls below 210 mN. Then net movement of both water and Na^+ is in the direction of lumen to blood, and net absorption of both continues until all the solution is absorbed.

Other osmotically active solutes have similar effects. In the oral glucose tolerance test, the solution drunk has a concentration of 2,000 mM or more. The first of this solution to reach the duodenum slows gastric emptying, but nevertheless a strongly hypertonic solution bathes the duodenal mucosa. Glucose is actively absorbed, but at the same time there is net movement of water from blood to lumen, and motility of the duodenum is strongly stimulated.

In normal circumstances, the contents of the intestine and colon below the duodenum are isotonic and remain so throughout digestion and absorption.

Along the whole of the small intestine net absorption of Na^+ is the resultant of two opposed unidirectional fluxes.

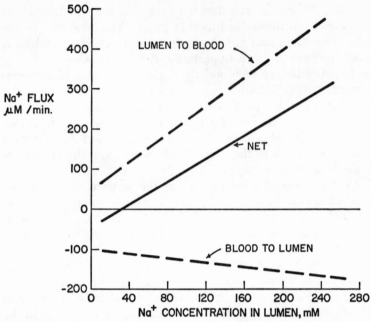

Fig. 10-2.—The net absorption of sodium from the small intestinal lumen as a function of the sodium concentration in the lumen. The unidirectional flux of sodium from blood to lumen is very slightly if at all affected by the luminal concentration, whereas the flux from lumen to blood, and consequently the net absorption, increases with increasing luminal concentration. (Adapted from Vaughan, B. E.: Am. J. Physiol. 198:1235, 1960.)

The flux from lumen to blood is chiefly caused by active transport of Na^+ by intestinal epithelial cells (Fig. 10-3). These cells pump Na^+ from their interior out their lateral border into the interstitial fluid. This keeps the concentration low inside the cells, and Na^+ diffuses from the luminal fluid into the cells. When glucose or galactose is present in luminal fluid, it is carried into the cells along with Na^+, the Na^+ diffusion gradient providing the energy for hexose transport. Na^+ pumped into the interstitial fluid between cells raises the osmotic pressure of the fluid confined there, and water flows through the cells in response to this osmotic gra-

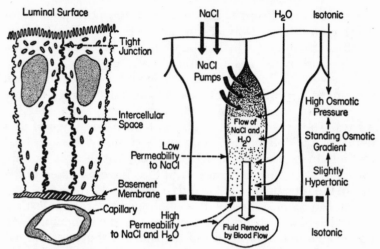

Fig. 10-3.—The mechanism of sodium and fluid transport by a layer of epithelial cells such as the mucosa of the intestine or of the gallbladder. In the intestinal epithelial cells, the sodium and chloride pumps are relatively independent of each other, but in the gallbladder, they are closely coupled. In each tissue, a standing osmotic gradient established between the lateral borders of the cells is responsible for water flow. (Adapted from Dietschy, J. M.: Gastroenterology 50:692, 1966.)

dient. Thus, absorption of water as well as absorption of glucose and galactose, is linked to active absorption of Na^+. The flux of Na^+ from blood to lumen is apparently accomplished by passive diffusion through water-filled pores.

Cl^- and HCO_3^- ions are in part passively absorbed along with Na^+. Cl^- can be actively absorbed as well, however. In the jejunum and ileum, this is accomplished by exchange of Cl^- for HCO_3^-, the latter ion being actively secreted. Therefore, the concentration of HCO_3^- in the lumen rises as the concentration of Cl^- falls, and chyme becomes more alkaline as it approaches the terminal ileum.

Potassium is absorbed by passive diffusion throughout the small intestine, and at equilibrium its concentration in intestinal contents ranges from 4 to 11 mN.

Most absorbed iron is recycled, but iron absorption must keep up with loss. Normal iron intake is 15 to 20 mg a day, most of it contained in hemoglobin and myoglobin. Heme iron is more easily absorbed than inorganic iron. A man absorbs about 0.5 to 1 mg a day. A menstruating woman, on account of her mild chronic anemia, absorbs more. The rate of absorption increases following hemorrhage, but not for 3 to 4 days. The reason for the delay is that the message of iron deficiency is delivered to cells in the crypts of the intestinal epithelium as they are being produced, but they do not absorb iron at an increased rate until they have migrated to the tips of the intestinal villi 3 to 4 days later.

Iron can be absorbed by the whole length of the intestine, but efficiency of absorption is greatest in the duodenum and upper jejunum. The nature of foodstuff in the intestine has more influence on iron absorption than does the form in which iron is fed. Iron bound to phosphoproteins and insoluble calcium phosphates is unavailable for absorption. Ascorbic acid reduces ferric to ferrous iron, the latter being more easily absorbed. By chelating iron, ascorbic acid releases iron bound to other compounds and makes it available for absorption.

Iron is actively absorbed into intestinal epithelial cells, and once inside the cells it enters two pools: one which can be absorbed into the plasma where iron combines with transferrin, and one which cannot be further absorbed (Fig. 10-4). When the load of iron in plasma is great, iron enters the intestinal epithelial cells, where it joins the pool of unabsorbable iron. As epithelial cells desquamate, the pool of unabsorbable iron is lost in the stool together with unabsorbable iron remaining in the intestinal lumen.

About two-thirds of the average daily intake of 1,000 mg of calcium is absorbed in the small intestine. In addition to dietary calcium, a large amount of calcium enters the intestinal lumen in bile, in desquamated cells and by diffusion from plasma. Calcium precipitated as phosphates or as soaps with fatty acids is not available for absorption. So much calcium

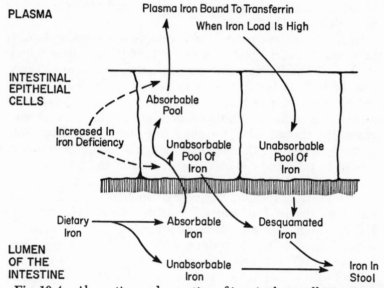

Fig. 10-4.—Absorption and excretion of iron in the small intestine.

may be lost with fatty acids in steatorrhea that negative calcium balance ensues.

Calcium is absorbed partly by diffusion but chiefly by active transport. Absorption is greatest in states of calcium deficiency, in young persons and in lactating women. Active transport is slightly increased by parathyroid hormone and greatly increased by vitamin D.

By the time intestinal contents reach the terminal ileum, digestion and absorption are complete, and a small volume of fluid containing residues is passed on to the colon. There are no good data on the volume actually delivered per day, but an imprecise guess can be made from a study of ileostomy fluid. This is the fluid collected from the opening of the ileum on the abdominal wall of patients who have had a colectomy. Shortly after an ileostomy is made, the rate of fluid loss is high, but eventually it declines. There is a conflict of opinion about the reason for the decline: some believe that patients learn to adjust their fluid intake, but others have found no relation between fluid intake and the volume lost through an ileostomy.

In representative patients, the volume of ileostomy fluid is between 380 and 1,500 ml a day. Normal fecal water output is about 100 ml a day. Na^+ concentration is 107 to 133 mN, and the mass of Na^+ lost is 44 to 197 mEq per day. This loss can be contrasted with normal fecal excretion of 0.5 to 5 mEq a day. This means that if ileostomy fluid is a measure of what enters the normal colon, the colon has a really significant role in conserving Na^+. The concentration of K^+ in ileostomy fluid is 4 to 11 mN, and the mass lost is 2 to 15 mEq a day. This loss is within the range of normal fecal excretion of K^+.

In perfusion studies, the normal human colon has been found to absorb Na^+ and Cl^- and to secrete HCO_3^-. The HCO_3^- reacts with organic acids produced by bacteria in the colon. Water is absorbed along with Na^+ at a maximum rate of about 2,000 ml a day. Consequently, the capacity of the colon to absorb, although limited, is well within the range of the burden imposed upon it in normal circumstances.

Diarrhea is defined as fecal water output greater than 500 ml a day. It may occur because the colon's capacity to absorb water and electrolytes is overwhelmed or because the colon's capacity to absorb is reduced. If osmotically active solutes are not absorbed in the small intestine, they carry water with them into the colon. Magnesium and sulfate ions are only slowly absorbed, and consequently ingestion of Epsom salts causes watery diarrhea. Lactose or trehalose may not be absorbed, because lactase or trehalase is absent from the brush border of the intestinal epithelial cells. If so, these sugars are fermented by intestinal flora, and the products cause osmotic diarrhea. The toxins of *Vibrio cholerae* and of some strains of *Escherichia coli* make the mucosa of the small intestine secrete enormous amounts of a fluid similar to an ultrafiltrate of plasma. Loss of this fluid results in potentially fatal dehydration.

The ability of the colon to absorb water and electrolytes may be reduced. Some bile acids inhibit colonic absorption of Na$^+$. Under normal circumstances, these bile acids do not reach the colon in significant amount, but when the ileum is diseased or resected, enough may enter the colon to cause diarrhea. Hydroxylated fatty acids such as ricinoleic acid, the active ingredient of castor oil, inhibit salt and water absorption. Unsaturated fatty acids are hydroxylated by intestinal and colonic bacteria, and when there are errors in fat digestion or absorption, hydroxylated fatty acids may be produced in amounts sufficient to cause diarrhea.

11. DIGESTION AND ABSORPTION OF CARBOHYDRATES

Carbohydrates make up 50 to 60% of the diet, 250 to 800 gm a day providing 1,000 to 2,500 kilocalories. The major carbohydrates are starches, glycogen and disaccharides. Starches contain long chains of glucose molecules linked by $\alpha-1,4$ glucosidic bonds. Branches originate at $\alpha-1,6$ bonds (Fig. 11-1). The chief disaccharides are sucrose, consisting of glucose and fructose, and lactose, consisting of glucose and galactose. There are many other carbohydrates in the diet including pentoses and disaccharides such as trehalose of mushrooms.

Hydrolysis of starch is catalyzed by salivary and pancreatic amylase. These enzymes attack only the $\alpha-1,4$ glucosidic bonds, and the products of hydrolysis of this bond are glucose, mannose (glucose-glucose), some trisaccharides and tetrasaccharides, all containing the $\alpha-1,4$ bond, and branched oligosaccharides, containing the $\alpha-1,6$ branching point.

Further hydrolysis is catalyzed by oligosaccharidases, which are confined to the brush border of the intestinal epithelial cells. Among these enzymes are maltase, which catalyzes the hydrolysis of the $\alpha-1,4$ bonds and isomaltase, which catalyzes hydrolysis of the $\alpha-1,6$ bonds. The product of hydrolysis is glucose.

The brush border also contains oligosaccharidases, which catalyze hydrolysis of other disaccharides. These include sucrase (or invertase), acting on sucrose to give glucose and fructose, and lactase, acting on lactose to give glucose and galactose.

Only monosaccharides are absorbed through the intestinal epithelial cells. Glucose and galactose are rapidly absorbed

83

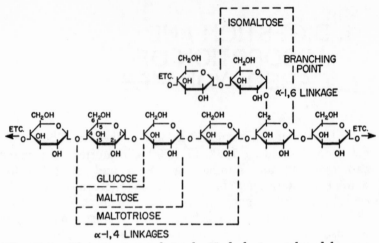

Fig. 11-1.—The structure of starch. Hydrolysis catalyzed by pancreatic amylase occurs at the α–1,4 linkage, and the products of hydrolysis are straight-chain oligosaccharides. Since pancreatic amylase does not catalyze hydrolysis of the α–1,6 branching point linkage, isomaltose is also a product of hydrolysis. Further hydrolysis is catalyzed by the maltases and the isomaltase of the brush border of the intestinal epithelial cells.

by active transport (Fig. 11-2). Xylose is absorbed by active transport as well. The other monosaccharides are absorbed by diffusion, and in the case of fructose, diffusion is facilitated. Since active transport is more rapid than diffusion, fructose accumulates in the intestinal lumen during digestion and absorption of sucrose. Later it disappears from the lumen as it diffuses into the epithelial cells.

Glucose and galactose share the same active transport mechanism, and consequently they compete for absorption. The transport process requires the presence of Na^+ in the lumen, and Na^+ is absorbed along with glucose and galactose. In order to cross the brush border, each Na^+ ion attaches itself to a carrier contained in the border. A glucose or a galactose molecule attaches itself to the same carrier and is transported into the cell along with the Na^+. The diffusion gradient for Na^+ from lumen to cell provides the energy necessary to transport the hexose into the cell.

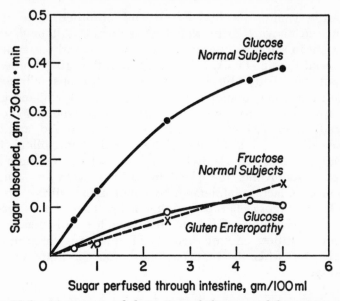

Fig. 11-2.—Mean rates of absorption of glucose and fructose in the jejunum of normal subjects and in patients with gluten enteropathy. The sugars were delivered in the concentrations shown by perfusion of a 30-cm length of jejunum at the rate of 20 ml per minute. Glucose absorption in the normal subjects appears to be by way of a saturable transport system. Fructose absorption in normal subjects appears to be by way of diffusion only. (Adapted from Holdsworth, C. D., and Dawson, A. M.: Gut 6:387, 1966.)

Some of the sugar accumulated in epithelial cells is used for energy, and some glucose is converted to glycerol to be used in synthesis of triglycerides and phospholipids. The rest escapes from the cells into interstitial fluid and plasma.

Active transport of glucose and galactose may be impaired in gluten enteropathy, a disease in which the luxuriant array of microvilli on intestinal epithelial cells is replaced by a few stubby ones, and the absorptive surface is drastically reduced.

The constituents of disaccharides cannot be absorbed until the disaccharide is hydrolyzed by a specific oligosaccharidase in the brush border. If the enzyme is absent, the disaccharide cannot be hydrolyzed and absorbed. The most common enzyme to be deficient is lactase. Although it is present in the brush border at birth, it frequently disappears at about the age of six. It is absent from the brush border of about 70% of adult black Americans, and it is absent from all adult Zambians. Sometimes it is deficient at birth, and fatal complications ensue unless the situation is promptly recognized. In absence of lactase, lactose is unabsorbed, and lactose fed in milk is fermented by intestinal microflora. Copious diarrhea follows. Switching from lactose to some other sugar circumvents the problem. Adults avoid the consequences of lactose intolerance by not drinking milk.

Other oligosaccharidase deficiencies have been recognized. A combined deficiency of isomaltase and sucrase has been identified, and a deficiency of trehalase, which results in diarrhea when its victims eat mushrooms, is a rare medical oddity.

12. DIGESTION AND ABSORPTION OF PROTEIN

⌈Under normal circumstances, most dietary protein is completely digested and absorbed. The protein of the stool is that contained in bacteria, desquamated cells and the mucoproteins of colonic secretions.⌋In states of maldigestion and malabsorption that cause steatorrhea, the loss of dietary protein closely parallels the loss of fat.

⌈Gastric digestion of protein is unnecessary; persons secreting no acid or pepsinogen digest and absorb protein adequately.

A very small amount of protein is absorbed intact through intestinal cells by the process of pinocytosis. An example of absorption of a large protein-containing molecule is furnished by the absorption of vitamin B_{12} and intrinsic factor in the terminal ileum. Intrinsic factor is a glycoprotein having a molecular weight between 45,000 and 50,000. It is secreted by the oxyntic cells of the stomach. It combines with vitamin B_{12} of the diet, and in doing so it forms dimers and trimers. The complex of vitamin B_{12} and intrinsic factor is absorbed intact into the epithelial cells of the ileum, and the vitamin is then transferred to other carriers in the plasma.⌋ The amount of other native proteins absorbed is nutritionally negligible, but allergic reactions to minute amounts of absorbed native proteins can be the basis of food sensitivities.

A large amount of endogenous protein is digested each day and absorbed as amino acids and polypeptides. There are three sources.

1. About 10 gm of protein enzymes is secreted into the intestinal tract each day. Most is digested and absorbed, but some pancreatic enzymes can be found in the stool.

2. There is a very rapid turnover of cells of the gastrointestinal epithelium, and the whole mucosa is replaced in about 3 days. This means that between 100 and 250 gm of mucosal cells is shed into the lumen each day, and these contribute 10 or more gm of protein to be digested and absorbed.
3. Under normal circumstances, a small amount of plasma proteins leaks into the digestive tract each day. The amount of albumin lost this way is about 1 to 4 gm a day, and a corresponding amount of other plasma proteins is probably lost into the gut as well. In states of protein-losing gastroenteropathy, an enormous amount of plasma proteins may be shed into the intestinal tract. In one instance, the proteins contained in 5 liters of plasma were found to be lost through the gastric mucosa in one day. Plasma proteins are digested along with other proteins, and their constituent amino acids are returned to the liver to be used for resynthesis of plasma proteins. In severe protein-losing states, however, the maximum synthetic ability of the liver is incapable of maintaining the normal plasma protein concentration, and plasma albumin concentration falls severely.

A large fraction of exogenous and endogenous protein is rapidly hydrolyzed in the duodenum and upper jejunum. Within 15 minutes after protein is emptied from the stomach, as much as 30 to 50% of it is broken down to amino acids and small polypeptides. The digestion products are almost entirely absorbed in the duodenum and jejunum. Complete digestion and absorption of 50 gm of protein in a meal, however, may take 4 to 6 hours, and this occurs throughout the whole small intestine.

Only 10% of the normal pancreatic capacity to secrete proteolytic enzymes is required for adequate hydrolysis of protein.

Amino acids and small polypeptides, those containing 2 or 3 amino acids, are absorbed by the epithelial cells at the tips of the intestinal villi. The polypeptides are absorbed more rapidly than are the free amino acids, and some amino acids accumulate in the lumen awaiting absorption. In the process of absorption, some polypeptides are hydrolyzed to their constituent amino acids by enzymes contained in the brush border, and the amino acids are then transported into the cells. Other polypeptides enter the cells intact and are hydrolyzed within the cells. A small fraction of the polypeptides that are absorbed appear intact in portal blood along with the amino acids.

Free amino acids are absorbed by at least three active transport systems.

1. Neutral amino acids are absorbed by one transport system, and as a result they compete with one another. The system is not absolutely specific for L-amino acids.
2. Basic amino acids are carried by a second system at a slower rate.
3. A third system transports proline and hydroxyproline.

The amino acid transport systems require the presence of Na^+ in the lumen, and the mechanism of transport may be similar to that of glucose and galactose.

Amino acids accumulate in intestinal epithelial cells during absorption. There is only a small rise in the concentration of amino acids in plasma during digestion of a meal, for amino acids are disposed of by tissues almost as quickly as they are released from intestinal epithelial cells.

In normal persons, some nitrogenous compounds escape into the colon, where they are attacked by colonic bacteria. Ammonia liberated by deamination is absorbed by diffusion into portal blood, and upon reaching the liver it is used for urea synthesis. In persons with liver disease or with portacaval shunt, the absorbed ammonia fails to be quickly removed by the liver, and the consequent rise in blood ammonia may cause encephalopathy.

In patients whose renal tubules are deficient in ability to reabsorb some amino acids, there may be a small but detectable deficiency in intestinal absorption as well. Persons with cystinuria do not absorb arginine or lysine as rapidly as do normal persons. Those whose renal tubules fail to reabsorb threonine and tryptophan have reduced intestinal absorption of tryptophan, and they may develop pellagralike symptoms that are correctable by feeding nicotinic acid. Unabsorbed amino acids are degraded in the colon to potentially toxic substances such as cadaverine and putrescine, and absorption of these through the colonic mucosa may lead to neurologic abnormalities.

Some persons in temperate climates have a collection of absorption defects called *nontropical sprue*—in infants, *celiac disease*. Because the disease can be traced to the effect of the wheat protein gluten upon the intestinal mucosa, it is also called *gluten enteropathy*. The mucosa is thinned, and the microvilli on the epithelial cells are clubbed and short. Absorption of carbohydrate, protein and fat are all impaired, and fermentation of food in the gut leads to bloating and diarrhea. The condition can be relieved by elimination of gluten from the diet.

13. BILE ACIDS

The major constituents of the bile are these:

1. A solution of sodium bicarbonate and chloride the secretion of which is independent of the secretion of bile acids, and which is controlled by the hormones secretin and cholecystokinin-pancreozymin. This component has been described in the section on pancreatic secretion (Fig. 13-1).
2. Bile acids, primary and secondary, almost entirely conjugated.
3. An electrolyte solution accompanying the bile acids the rate of secretion of which is governed by the rate of secretion of bile acids.
4. Lecithin.
5. Cholesterol, which, together with lecithin and bile acids, is chiefly contained in micelles but which may be in the form of microcrystals in hepatic bile.
6. Bile pigments, chiefly bilirubin conjugated with glucuronic acid.
7. Some protein.
8. Many compounds metabolized and secreted by the liver: detoxified drugs, phenolsulfonphthalein (PSP) in conjugated form, etc., etc.

Bile acids are steroid derivatives of cholesterol (Fig. 13-2). Their chemical features are the steroid nucleus to which 3, 2 or 1 hydroxyl groups are attached and a short side chain ending in a carbonyl group. Primary bile acids are those which are synthesized by the liver; these are cholic acid with three hydroxyl groups and chenodeoxycholic with two hydroxyl groups. Primary bile acids may be modified by bacterial action in the small intestine and colon, and the prod-

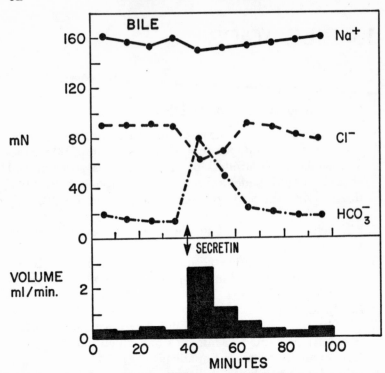

Fig. 13-1.—The composition of the bile acid-independent fraction of human bile. Secretion of an isotonic, bicarbonate-rich fluid is stimulated by secretin administered intravenously or released from the duodenal mucosa. (Adapted from Waitman, A. M., et al.: Gastroenterology 56:286, 1969.)

ucts are called secondary bile acids. The most important secondary bile acids are deoxycholic acid, derived from cholic acid which have two hydroxyl groups, and lithocholic acid, which has only one hydroxyl group. When secondary bile acids are absorbed and returned to the liver in portal blood, they are secreted into bile along with the primary bile acids.

The liver conjugates bile acids with glycine or taurine by forming a peptide bond between the carbonyl group on the

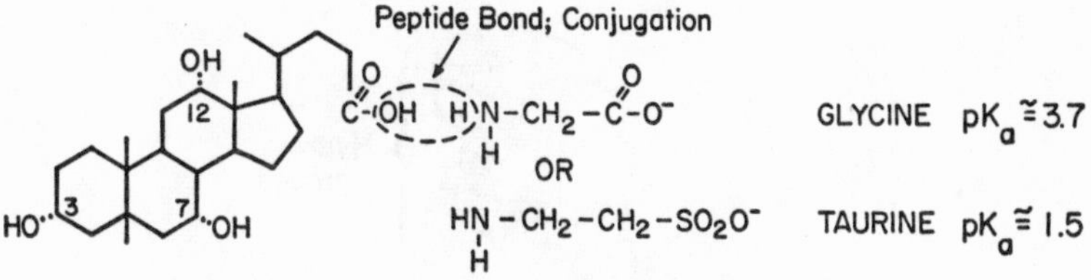

CHOLIC ACID (Primary Bile Acid)

NO – OH on 12 = Chenodeoxycholic Acid (primary)

NO – OH on 7 = Deoxycholic Acid (secondary)

NO – OH on 7 and 12 = Lithocholic Acid (secondary)

Fig. 13-2.—The structure of the common bile acids. The bile acid is conjugated with either glycine or taurine by elimination of water to form a peptide bond. The approximate ionization constants of glycocholic and taurocholic acids are given on the right. (Adapted from Hofmann, A. F.: Gastroenterology 48:484, 1965.)

side chain of the bile acid and the amino group of glycine or taurine. If the bile acid is cholic acid, the conjugates are called glycocholic and taurocholic acid, respectively. Conjugates of other bile acids have corresponding names. In man, the ratio of glycoconjugates to tauroconjugates is about 3 to 1. The reason is that taurine is in relatively short supply, whereas that of glycine is unlimited. If taurine is fed, more tauroconjugates are secreted. If taurine is lost by failure to reabsorb conjugated bile acids, the ratio of glycoconjugates to tauroconjugates may rise to 20 to 1. Taurine is a stronger acid than glycine, and consequently a greater fraction of tauroconjugates is ionized at the pH of intestinal contents.

The total amount of bile acids, primary and secondary, conjugated and unconjugated, in the body at any one time is called the *bile acid pool*, and in a normal man the pool size is about 2 to 4 gm.

A conjugated bile acid is a flat molecule, water-soluble on one side and fat-soluble on the other (Fig. 13-3). The hydrophilic hydroxyl groups, the peptide bond and the terminal group of glycine or taurine project to one side, and the other side is made up of hydrophobic groups. Consequently, bile

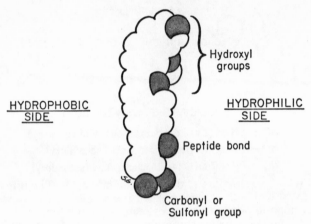

Fig. 13-3.—A conjugated cholic acid molecule (side-on view). (Adapted from D. M. Small.)

acids accumulate at oil-water interfaces, and they stabilize the interface.

All bile acids, whether conjugated or not, returning to the liver in portal blood are promptly secreted into the bile (Fig. 13-4, **A**). Unconjugated bile acids are conjugated before they are secreted. A small amount of unconjugated bile acids is present in bile only when massive amounts of unconjugated bile acids reach the liver. Bile acids are always secreted in conjunction with lecithin and cholesterol. These three compounds aggregate in micelles. Both bile acids and lecithin are polar molecules; in a micelle, their fat-soluble portions form a hydrophobic core and their water-soluble portions form a hydrophilic shell. Cholesterol, which is almost totally insoluble in water, dissolves in the hydrophobic core of the micelle.

Bile is a neutral solution, and the bile acids are ionized. Their negative charges are balanced by cations, chiefly sodium, and these form a counterion shell around the micelles. Osmotic pressure is determined by the total number of particles in solution, and bile is isotonic with plasma. Because the organic constituents of bile, together with the cations, aggregate in micelles, the total number of particles in solution is far less than the total number of particles determined by chemical analysis. The osmotic coefficient of Na^+ in bile is only about 0.3. Any increase in the rate of bile acid secretion is accompanied by an increase in the volume flow of bile. This is called the *choleretic* effect of bile acids.

During the interdigestive period, the gallbladder is relaxed, and resistance of the sphincter separating the common bile duct from the duodenum is relatively high (Fig. 13-4, **B**). Only a small fraction of the bile secreted by the liver flows into the intestine; the rest is diverted to the gallbladder. The mucosa of the gallbladder actively reabsorbs Na^+, Cl^- and HCO_3^- ions from gallbladder bile, and as these are reabsorbed, water passively follows. The result is that at the end of the interdigestive period the gallbladder contains 40 to 60 ml of an isotonic solution in which the bile acid-lecithin-cholesterol micelles are highly concentrated. Bile acids may be 5 to 10 times more concentrated in gallbladder bile than in hepatic bile. Because bile pigments are concentrated as well, gallbladder bile is almost black (hence, *melancholia*, a disorder of black bile).

Soon after the beginning of a meal, the first chyme to reach the duodenum stimulates release of cholecystokinin from the duodenal mucosa. The hormone causes the gallbladder to contract, and over the next 20 or so minutes the gallbladder's store of concentrated bile enters the duodenum (Fig. 13-4, **C**). Bile acids travel down the intestine with the chyme and are reabsorbed, only to be secreted once more by the liver. The result is that throughout the digestion and absorption of a meal, the concentration of bile acids in intestinal contents is adequate for their physiologic function.

The functions of bile acids in digestion and absorption are the following:

1. Bile acids permit emulsification of fat by reducing the tension of the oil-water interface.
2. Bile acids prevent denaturation of pancreatic lipase as it leaves the surface of emulsified fat droplets.

3. Bile acids, together with 2-monoglycerides, which are the products of fat hydrolysis, form micelles which then dissolve cholesterol, free fatty acids and fat-soluble vitamins.

These functions are discussed in the next section.

During digestion and absorption, bile acids travel with chyme along the intestine, and they are almost entirely absorbed into portal blood. Upon reaching the liver, they are again secreted into bile, and consequently they circulate from liver to intestine to liver and to intestine again. This process is called the *enterohepatic circulation* of bile acids.

Absorption of bile acids is both active and passive.

Active absorption occurs only in the last part of the ileum, and only ionized, conjugated bile acids are actively absorbed. The absorption process is so efficient that only a small percentage of conjugated bile acids, less than 5%, reaching the terminal ileum escapes into the colon. In normal man, absorption of conjugated bile acids in the ileum accounts for most of the absorptive phase of the enterohepatic circulation.

Passive absorption of bile acids occurs in two ways, both depending on fat-solubility.

1. Un-ionized conjugated bile acids are more fat-soluble than are ionized conjugated bile acids. At the pH of intestinal contents, all tauroconjugates are ionized, and they are not passively absorbed. Because glycine is a weaker acid than taurine, a fraction of the glycoconjugates is un-ionized and fat-soluble. A small amount of glycoconjugates is therefore absorbed by passive diffusion through the lipid membrane of the intestinal mucosal cells.

2. Bile acids are attacked by intestinal bacteria. Some bile acids are dehydroxylated; cholic acid loses one hydroxyl group to become deoxycholic acid, and bile acids with two hydroxyl groups lose one to become lithocholic acid. These secondary bile acids, after being absorbed, become part of the bile acid pool. Bacteria also deconjugate acids by hydrolyzing the peptide bond between the acid and glycine or taurine. Deconjugated bile acids are more fat-

soluble than are conjugated bile acids, and therefore some deconjugated bile acids are passively absorbed. Once they reach the liver, they are reconjugated and secreted in the bile.

As the result of bacterial action, bile acids are deconjugated many times as they circulate. In addition, primary bile acids are converted to secondary bile acids which enter the bile acid pool. One consequence of excessive bacterial action is that the concentration of conjugated bile acids in intestinal contents is reduced, and the concentration of unconjugated bile acids is increased. Since only conjugated bile acids can be actively reabsorbed and since unconjugated bile acids are poorly absorbed only by passive diffusion, a larger fraction of circulating bile acids escapes into the colon. The size of the bile acid pool is reduced, and the fraction of secondary bile acids in the pool is increased.

The rate of enterohepatic circulation is determined by the digestive cycle. During the interdigestive period, most bile acids are sequestered in the gallbladder, and the rate of circulation is low. During digestion of a meal, the gallbladder empties, and bile acids circulate two or more times. In a man eating a low-calorie diet, the pool may circulate 3 to 6 times a day, but on a high-calorie diet, the pool circulates 5 to 14 times a day. If the pool contains a total of 3 gm of bile acids, the man has the use of 9 to 42 gm of bile acids for the purposes of daily digestion.

Each day about 500 mg of bile acids fail to be absorbed in the small intestine. This amount escapes into the colon, where some of it is degraded by bacteria, some absorbed and some excreted in the stool. Bile acids lost are replaced by new bile acids synthesized in the liver, and pool size remains constant. If the terminal ileum is diseased or has been surgically removed, the enterohepatic circulation is broken (Fig. 13-4, **D**). A large fraction of conjugated bile acids is not absorbed but passes into the colon.

The rate at which the liver synthesizes bile acids is governed by the rate at which bile acids return to the liver in

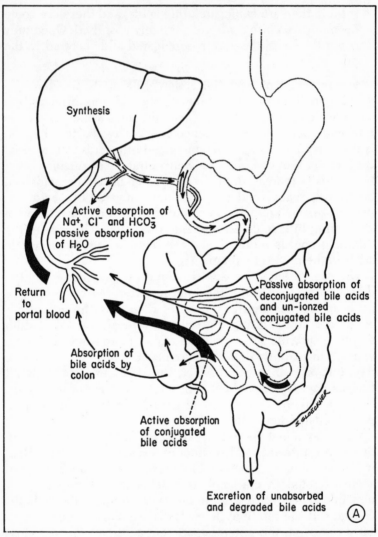

Fig. 13-4.—The enterohepatic circulation of bile acids. **A,** the complete system: synthesis in the liver, secretion of newly synthesized and reabsorbed bile acids, storage in the gallbladder, passive and active absorption in the small intestine, escape of a small fraction into the colon, passive absorption in the colon and excretion into the feces, and return of absorbed bile acids to the liver in the portal blood. (Continued.)

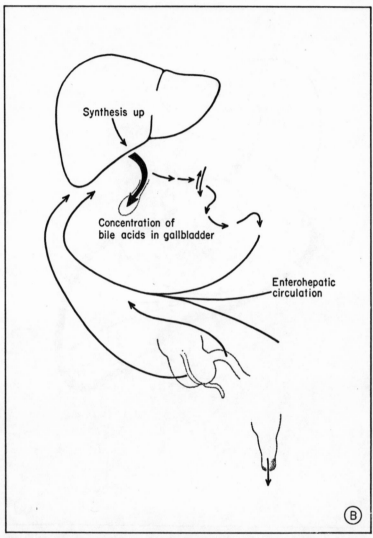

Fig. 13-4 (cont.).—B, the interdigestive phase: secretion and storage in the gallbladder, minimal bile acids in intestine with minimal absorption and return to the liver, increased synthesis as a result of reduced negative feedback control. (Continued.)

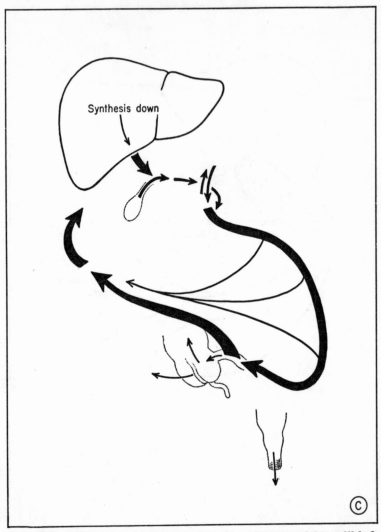

Fig. 13-4 (cont.).—C, the digestive phase: emptying of the gallbladder, large amounts of bile acids in the intestine with rapid absorption, rapid secretion of reabsorbed bile acids, reduced synthesis on account of increased negative feedback control. (Continued.)

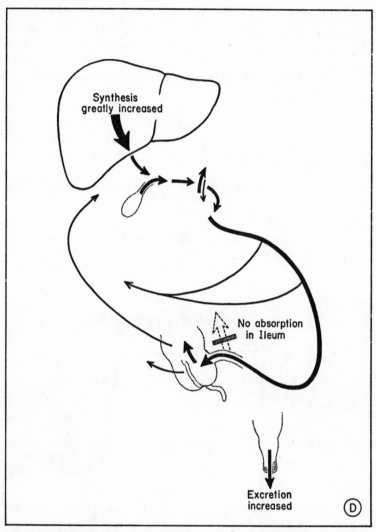

Fig. 13-4 (cont.).—D, interruption of active reabsorption in the terminal ileum: decreased or absent active reabsorption, decreased return to liver with greatly increased synthesis, increased escape of bile acids into colon and feces.

portal blood (Fig. 13-5). As the rate of return falls, the rate of synthesis rises steeply to a maximum. If about 20% of the bile acids fail to return, synthesis is adequate to keep pool size constant at the normal level. If more than 20% is lost, pool size falls, because the maximum rate of replacement has been reached.

When the enterohepatic circulation has been broken, there is diurnal variation in the amount of bile acids available. Late at night, when the intestine is empty, the bile acids synthesized by the liver are collected and stored in the gallbladder.

Fig. 13-5.—The feedback control of bile acid synthesis by the liver. Under normal circumstances, most of the bile acids return to the liver during each cycle of the enterohepatic circulation. The rate of bile acid synthesis is just sufficient to replace the bile acids lost, and bile acid pool size is maintained. If the rate of return of bile acids to the liver falls, the rate of bile acid synthesis rises sharply. Bile acid pool size remains at the normal level until a little more than 20% of the bile acids is lost during each cycle. Then the maximum rate of synthesis is reached, and with further loss of bile acids the pool size falls.

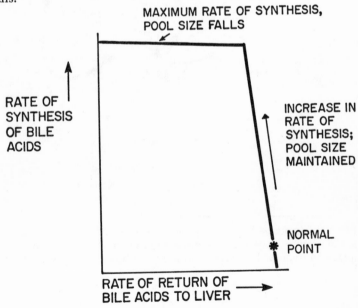

MAXIMUM RATE OF SYNTHESIS, POOL SIZE FALLS

RATE OF SYNTHESIS OF BILE ACIDS

INCREASE IN RATE OF SYNTHESIS; POOL SIZE MAINTAINED

NORMAL POINT

RATE OF RETURN OF BILE ACIDS TO LIVER

As the first meal is eaten, these bile acids are emptied into the duodenum, and they may be adequate for the digestion of the meal. If they are not reabsorbed, however, only the bile acids immediately synthesized are available for the rest of the day. This amount is usually inadequate for normal digestion of fat.

Cholesterol gallstones form in the gallbladder when minute crystals of cholesterol contained in bile aggregate. Cholesterol is carried in micellar solution in bile secreted by the liver. Since micelles are composed of bile acids, lecithin and cholesterol, the amount of cholesterol that can be carried in micellar solution depends upon the amount of bile acids and lecithin. If the proportion of bile acids and lecithin is high compared with the amount of cholesterol, the cholesterol secreted by the liver is held in micellar solution. When the bile is concentrated in the gallbladder, the proportion of bile acids and lecithin to cholesterol remains unchanged; cholesterol continues to be held in micellar solution. However, if the concentration at which cholesterol is secreted by the liver is greater than that which can be carried by the bile acids and lecithin secreted at the same time, microcrystals of cholesterol form in the bile. When bile containing these crystals is concentrated in the gallbladder, the crystals may coalesce into stones.

Some persons frequently or continuously secrete bile containing more cholesterol than can be held in micellar solution; they have microcrystals of cholesterol in their hepatic bile and cholesterol stones in their gallbladder. We do not know why their livers secrete bile containing a high proportion of cholesterol. One factor may be that they have small amounts of bile acids to secrete along with cholesterol. Persons with cholesterol gallstones have a total bile acid pool smaller than that of normal persons. Feeding such persons chenodeoxycholic acid increases their bile acid pool, reduces the tendency of the liver to secrete bile supersaturated with cholesterol and promotes dissolution of stones already present in the gallbladder.

Persons without gallstones may secrete bile containing

microcrystals of cholesterol some parts of the day and bile without microcrystals the rest of the time. Why stones do not form in their gallbladder is unknown.

14. DIGESTION AND ABSORPTION OF FAT

Because fat contains 9 kilocalories per gm, it is a major source of dietary calories. Intake of fat ranges from 12 gm a day (108 kilocalories) in the poorest populations to 150 gm a day (1,350 kilocalories) in the most self-indulgent.

The most important physical property of fat is that fats are poorly soluble in water. For this reason, fat is the major structural component of the body. Fat forms the membranes of cells, and without membranes the body would thaw and resolve itself into a dew.

Insolubility of fat presents a physiologic problem: How can fat be transferred from food through the aqueous media of chyme, cell cytoplasm, interstitial fluid, lymph and blood to the organs in which it can be used for energy and structure? To solve this problem, the body carries fat through a complex series of physical and chemical transformations (Fig. 14-1). Fat is emulsified and then hydrolyzed to free fatty acids and 2-monoglycerides. These are held in micelles in the lumen until they can be absorbed. Inside the intestinal epithelial cell, most of the free fatty acids and 2-monoglycerides are resynthesized into triglycerides and phospholipids. These are aggregated into small droplets, the chylomicrons, on whose surface a β-lipoprotein is spread, and the chylomicrons are extruded from the cells into the interstitial fluid. They find their way into the lymph and are eventually delivered to the blood. Each step is subject to error, and error at one or another step results in maldigestion or malabsorption of fat.

Triglycerides are the most numerous and important fats in the diet. They are esters of glycerol and three fatty acids. All the fatty acids have an even number of carbon atoms, and

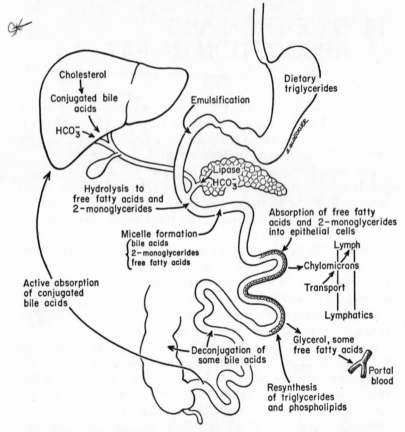

Fig. 14-1.—The processes of fat digestion and absorption.

most of the fatty acids have long chains, 16 carbon atoms in the case of palmitic acid and 18 in stearic acid. Fatty acids may have one or more double bonds. The fatty acid with 18 carbons and one double bond is oleic acid, and that with two double bonds is linoleic acid.

The longer the chain length of the fatty acids in triglycerides, the higher is the melting point. Triglycerides containing saturated fatty acids have higher melting points than those containing unsaturated acids. The melting point has an important influence upon the digestibility of triglycerides. Triolein is an oil at body temperature, and it is 96% digested and absorbed. On the other hand, tripalmitin is solid at body temperature, and it is only 12% digested and absorbed.

Very few triglycerides containing short-chain fatty acids are eaten in the normal diet. The only common source of short-chain fatty acids is milk and butter, and less than 10% of the fatty acids in butter have a chain length shorter than 10 carbons long. There are almost no medium-length fatty acids 10, 12 or 14 carbons long in the normal diet.

Triglycerides of short-chain and medium-chain fatty acids are more water-soluble than those of long-chain fatty acids, and they can be more completely digested and absorbed when the digestion of triglycerides of long-chain fatty acids is impaired by errors.

A fatty acid esterified at either of the terminal hydroxyl groups of glycerol is said to occupy the 1 or the 1' position. A fatty acid esterified at the middle hydroxyl group is said to occupy the 2 position (Fig. 14-2).

Lecithin is a phospholipid; fatty acids are esterified at the 1 and 2 positions of glycerol, and phosphoric acid is esterified at the 1' position (Fig. 14-3). An organic base such as choline or ethanolamine is bound in an ester bond to the phosphoric acid. Consequently, the lecithin molecule is polar: the fatty acid end is hydrophobic and the phosphoric

Fig. 14-2.—The hydrolysis of a triglyceride catalyzed by pancreatic lipase. The reaction goes completely to the right for the reason that free fatty acids are ionized at the pH of intestinal contents. Pancreatic lipase catalyzes the formation of the ester bond only with un-ionized fatty acids.

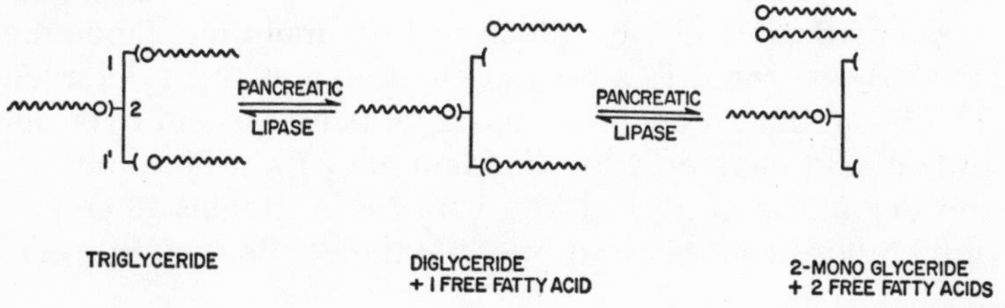

TRIGLYCERIDE   DIGLYCERIDE   2-MONO GLYCERIDE
                      + I FREE FATTY ACID + 2 FREE FATTY ACIDS

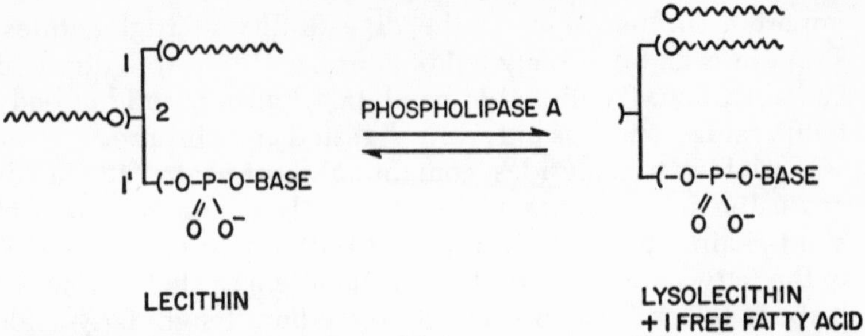

Fig. 14-3.—The hydrolysis of lecithin of food and bile to lysolecithin and a free fatty acid catalyzed by phospholipase A of pancreatic juice.

acid and base make the other end hydrophilic. There is some lecithin in the normal diet, and there is much lecithin in the bile. Lecithin from both sources is digested and absorbed. The amount of lecithin digested has no influence upon the amount secreted in the bile.

The enzyme phospholipase A contained in pancreatic juice catalyzes the hydrolysis of the ester bond of lecithin at the 2 position. The product is lysolecithin: glycerol esterified to a fatty acid at the 1 position and to the phosphoric acid-base compound at the 1' position. The molecule is highly polar, and there is a high concentration of it in duodenal chyme during normal digestion.

The lipase secreted by the gastric mucosa is most active in catalyzing the hydrolysis of triglycerides containing short-chain fatty acids, and it is inactive in an acid medium. It is probably unimportant in adults. Acid in the stomach itself promotes the hydrolysis of some ester bonds, and it tends to break emulsions of fat.

Emulsification of most dietary fat occurs in the duodenum. The process requires a neutral environment and a detergent. The bicarbonate-containing juices of pancreas and liver neutralize acid chyme in the duodenum. Within 20 minutes of the beginning of a meal, the gallbladder begins to contract and empties concentrated bile into the duodenum. Bile acids and lecithin thereupon emulsify fat, forming a stable emulsion of droplets 0.5 to 1 micron in diameter.

Lipase is mixed with duodenal chyme at the same time. Pancreatic lipase has two important properties: (1) it acts as a catalyst only when it is spread on the surface of an emulsified fat droplet, and (2) it is almost entirely specific in catalyzing the hydrolysis of the ester bonds at the 1 and 1' positions. Consequently, the products of lipolytic action are free fatty acids and 2-monoglycerides (Fig. 14-4). These are the major

Fig. 14-4.—The course of hydrolysis of a triglyceride catalyzed by pancreatic lipase. There is a transient appearance of diglycerides, and the final products of rapid enzymatic hydrolysis are 2-monoglycerides and free fatty acids. Then there is a slow decline in the amount of 2-monoglycerides and a corresponding slow appearance of additional free fatty acids as the ester bond on the 2-position of glycerol is hydrolyzed. At the same time, there is the slow appearance of free glycerol. (After Mattson, F. H.)

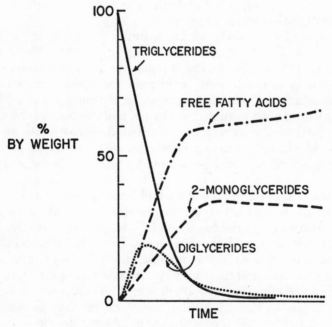

fat digestion products, which are subsequently absorbed, and during absorption at least 75% of the ester bonds between fatty acids and the hydroxyl groups at the 2 position are preserved intact. These intact ester bonds appear in triglycerides and phospholipids in the chylomicrons of intestinal lymph.

The ester bond in triglycerides is a low-energy bond, and re-esterification is relatively easy. The equilibrium between glycerides and free fatty acids is shifted in the direction of complete hydrolysis by the ionization of free fatty acids at the pH of chyme. Because the free fatty acids are ionized, they cannot reunite with glycerol; only un-ionized fatty acids can esterify.

Another consequence of the fact that the ester bond is a low-energy bond is some hydrolysis at the 2 position. This produces some free glycerol which, being water-soluble, is absorbed by diffusion and passes into the portal blood. Upon reaching the liver, it is metabolized and plays no further part in fat digestion and absorption.

The major products of fat hydrolysis remaining in the chyme are free fatty acids and 2-monoglycerides. These are only slightly soluble in the aqueous medium of chyme. Hydrolysis is so much more rapid than absorption that, unless some additional provision were made, the products of hydrolysis would soon saturate the chyme, and the bulk of the free fatty acids and 2-monoglycerides would separate into an oil or solid phase. Once separated, the free fatty acids and 2-monoglycerides would be essentially unavailable for absorption.

Separation of free fatty acids and 2-monoglycerides into an oil phase is prevented by micelle formation. Micelles are aggregates of fat molecules and bile acids. Fats form the hydrophobic core, and the bile acids, being polar molecules, cover the surface of the micelle, their hydrophobic side facing the core and their hydrophilic side facing the aqueous medium. Each micelle is about 4 to 6 millimicrons in diameter. It has a volume a million times smaller than an emulsified fat droplet, and it contains about 20 fat molecules.

The initial constituents of micelles are bile acids and 2-monoglycerides. Two characteristics of bile acids determine whether micelles are formed: their Krafft point and their Critical Micellar Concentration.

The Krafft point is the temperature below which a particular bile acid will not form a micelle. Most bile acids have Krafft points well below body temperature and are therefore capable of forming micelles in the intestinal lumen. The secondary bile acid, lithocholic acid, has a high Krafft point and is incapable of forming micelles at body temperature.

The Critical Micellar Concentration is the minimal concentration of a particular bile acid required for micelle formation (Fig. 14-5). When the concentration of the bile acid is at or

Fig. 14-5.—Relationship between bile acid concentration and the quantity of mono-olein brought into micellar solution at pH 6.3, 37°C and [Na$^+$] of 150 mN. The critical micellar concentration (CMC) of the conjugated cholic acid is 0.25 mM, a concentration well below that usually occurring in intestinal contents. The CMC for the unconjugated bile acid is 4 mM. (Adapted from Hofmann, A. F., and Borgstrom, B.: Fed. Proc. 21:43, 1962.)

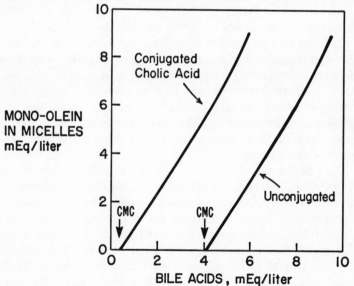

above its Critical Micellar Concentration, the bile acid and 2-monoglycerides aggregate as micelles, and with increasing bile acid concentration more 2-monoglycerides are carried in micelles. The Critical Micellar Concentration of conjugated bile acids is well below the usual concentration of those bile acids in chyme, and micelles easily form as 2-monoglycerides are produced by the hydrolysis of triglycerides. The Critical Micellar Concentration of unconjugated bile acids is much higher than that of conjugated acids. Consequently, when a considerable fraction of bile acids is deconjugated in the intestinal lumen by bacterial action, micelle formation is reduced. When deconjugated bile acids predominate, fewer conjugated bile acids are present to form micelles, and the deconjugated bile acids cannot form micelles on account of their high Critical Micellar Concentration.

Once micelles are formed by bile acids and 2-monoglycerides, they dissolve other fat-soluble compounds. Quantitatively, the most important of these are the free fatty acids. Fat-soluble vitamins are carried in micelles, and this is the reason bile acids are essential for absorption of vitamin K. Cholesterol and esters of cholesterol, compounds which are almost totally insoluble in water, are dissolved in micelles.

The constituent molecules of micelles move back and forth between micelles and solution with great rapidity. The mean residence time of a particular molecule in a micelle is only 10 milliseconds. This means that as free fatty acids and 2-monoglycerides are absorbed from solution into the epithelial cells, the aqueous phase of chyme is kept saturated by movement of free fatty acids and 2-monoglycerides from micelles to solution.

Most of the free fatty acids and 2-monoglycerides are absorbed from solution by diffusion through the membrane of the intestinal epithelial cells. Once inside the cells, they are resynthesized into triglycerides and phospholipids, and their concentration in the apical part of the epithelial cells is low. This provides a concentration gradient down which the free fatty acids and 2-monoglycerides can diffuse from the luminal solution into the cells. As they leave the solution by entering the cells, the solution is quickly resaturated with free fatty acids and 2-monoglycerides from micelles. Thus micelles perform a holding operation in fat digestion and absorption, preventing the products of lipolysis from separating into an oil or solid phase and serving as a reservoir of lipolytic products so that the aqueous phase is always saturated (Fig. 14-6).

Most free fatty acids and 2-monoglycerides are absorbed in the duodenum and upper jejunum. Most bile acids are absorbed in the terminal ileum. A little cholesterol is absorbed throughout the length of the intestine, but most escapes into the colon. Therefore, the constituents of micelles are separated as the chyme passes down the intestine; their content of free fatty acids and 2-monoglycerides diminishes and their concentration of cholesterol rises.

Once inside the intestinal epithelial cells, the 2-monoglycerides undergo little or no further hydrolysis (Fig. 14-7). They are resynthesized into triglycerides. Some of the free fatty acids are combined with glycerophosphate derived from glucose to form phospholipids, and some of the free fatty

Fig. 14-6.—The importance of micelle formation during the digestion and absorption of fat.

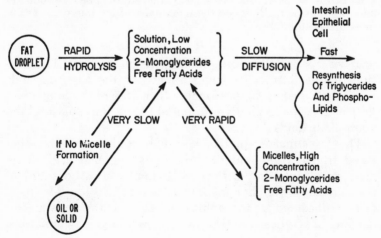

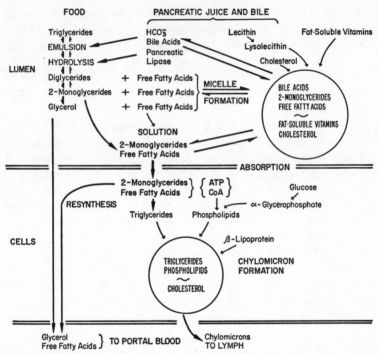

Fig. 14-7.—The process of fat digestion and absorption. It is possible that for some substances such as cholesterol to be absorbed the micelles must actually touch the membrane of the microvilli of the intestinal epithelial cells.

acids escape into the portal blood to reach the liver. The fraction of free fatty acids going directly to the liver depends upon their chain length. No more than 15% of the long-chain fatty acids goes to the liver in portal blood, but almost all the short-chain acids do.

The intestinal epithelial cells collect the newly synthesized triglycerides and phospholipids, together with some cholesterol, into droplets called chylomicrons. About 10% of the surface of the chylomicrons is covered with a β-lipoprotein synthesized by the ephithelial cells. If the cells are incapable of synthesizing this protein, only large chylomicrons, or none at all, are formed. The cells extrude the chylomicrons from their lateral borders into the interstitial fluid.

Chylomicrons are absorbed into the intestinal lymphatic vessels and transported through the thoracic duct into the blood. Chylomicrons enter the lymphatic vessels rather than the intestinal capillaries, because the fenestrations of the lymphatic vessels are open whereas those of the capillaries are closed by a basement membrane.

Under normal circumstances, there is less than 6 gm of fat in the stool each day, and the amount of fat in the stool is independent of the diet. Most normal fecal fat is contained in bacterial cells. If there are errors in fat digestion and absorption, fat escapes into the stool. Excess fat in the stool, or steatorrhea, is usually defined as more than 6 gm a day.

In steatorrhea, the amount of fecal fat is roughly proportional to the amount of dietary fat (Fig. 14-8), but the fat in

Fig. 14-8.—Relationship between dietary fat and total fecal fat in man. *Left*, in complete absence of bile. The slope of the regression line indicates that on the average 46% of dietary fat appears in the stool in the complete absence of bile. *Right*, in complete absence of pancreatic juice. The slope of the regression line indicates that on the average 68% of dietary fat appears in the stool in the complete absence of pancreatic juice. (Adapted from Annegers, J. H.: Q. Bull. Northwestern Univ. Med. School 23:198, 1949.)

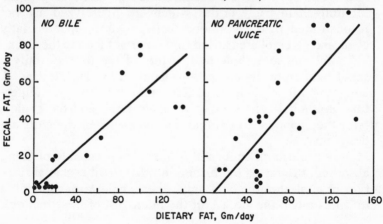

the stool is different in some respects from the fat in the diet. Most fecal fat is in the form of free fatty acids; triglycerides of the diet are hydrolyzed and the glycerol absorbed. All the short-chain and most of the medium-chain fatty acids of the diet are absent, for they too are absorbed. Abnormal fats are present. Intestinal bacteria add hydroxyl groups to the double bonds of unsaturated fatty acids. One of the products is ricinoleic acid, the active ingredient of castor oil, and the presence of this hydroxylated fatty acid accounts, in part, for the diarrhea which often accompanies steatorrhea.

Errors in fat digestion and absorption leading to steatorrhea are summarized in Table 14-1.

The remarkable fact is that if 40 to 60% of dietary fat fails to be absorbed, then 60 to 40% of dietary fat actually is absorbed in diseased states. Processes responsible are poorly understood.

Medium-chain triglycerides are not abundant in nature, but medium-chain triglycerides commercially synthesized are often used to supply calories to persons with errors in fat digestion and absorption. These triglycerides, containing fatty acids 6 to 12 carbons long, are rapidly and completely hydrolyzed in the intestinal lumen, and they can also be hydrolyzed within the epithelial cells. Hydrolysis and micelle formation, however, are not necessary for their absorption. Medium-chain triglycerides which are hydrolyzed are not reconstituted in the epithelial cells; their component fatty acids and glycerol are carried in the portal blood to the liver.

Cholesterol and cholesterol esters of the diet are only a small fraction of the cholesterol handled by the intestine. A large amount of cholesterol is contained in the bile. Intestinal epithelial cells synthesize cholesterol, and the cholesterol they contain also enters the lumen when the cells are desquamated.

Only cholesterol contained in micelles is capable of being absorbed; the rest is too insoluble. It is possible that absorption of cholesterol occurs only when cholesterol-containing micelles touch the surface of the microvilli of intestinal epi-

TABLE 14-1.   ERRORS IN FAT DIGESTION AND ABSORPTION
LEADING TO STEATORRHEA

STEP	PHYSIOLOGICAL DISTURBANCE	DISEASE STATE
Emulsification of triglycerides	Impaired emulsification	Deficiency of conjugated bile acids; excessive acidity of intestinal chyme
Hydrolysis	Pancreatic lipase deficiency	Pancreatic disease
	Absolute or relative bicarbonate deficiency	Pancreatic disease or gastric hypersecretion
Formation of micelles	Conjugated bile acid deficiency; absolute or relative bicarbonate deficiency	Biliary fistula or obstruction; ileal disease or resection; bacterial deconjugation; pancreatic disease or gastric hypersecretion
Absorption of 2-monoglycerides	Decreased cell uptake; reduction in cell number, activity or surface area	Intestinal resection or bypass; tropical sprue or gluten enteropathy
	Cells saturated with fatty acids and monoglycerides	Failure of triglyceride synthesis, chylomicron formation or transport
	Decreased contact time	Increased transit time
Chylomicron formation	Deficiency in chylomicron formation	A-β-lipoproteinemia
Transport of chylomicrons from cells via lymph to blood	Lymphatic obstruction or lymphangiectasia	Lymphosarcoma; intestinal lipodystrophy; protein-losing enteropathy

thelial cells. Then the cholesterol is transfered to the lipid
layer encasing the microvilli.

15. GAS IN THE GUT

The four sources of intestinal gas are swallowing, fermentation, neutralization and diffusion.

1. *Swallowing.* Air is swallowed with food and drink in very variable amounts. Some persons swallow so much while eating that they become uncomfortably bloated, whereas the air swallowed by others accounts for less than a third of the gas in the gut. Air is also swallowed in frothy saliva, and much may be swallowed during copious salivation accompanying nausea. Oxygen is removed from swallowed air by the flora of the gut, and the residual nitrogen is excreted as flatus.

2. *Fermentation.* Under normal circumstances, there are few bacteria in the small intestine. Gastric contents are partially sterilized by acid secreted by the stomach, and chyme moves rapidly from duodenum to colon. When chyme moves with abnormal slowness or when food residues are sequestered in blind loops or diverticuli, bacteria multiply. Then the small intestine becomes bloated with gas that is produced by fermentation. Fermentation and gas production normally occur in the slowly moving contents of the colon. Colonic flora of all persons produce variable amounts of hydrogen gas. About 14% is absorbed and excreted through the lungs; the rest escapes in flatus. About a third of the population produces a large amount of methane in the colon, but the other two-thirds does not. Some methane is also absorbed into the blood, and its mass-spectrometric measurement in breath is a means by which its production in the gut can be followed. Methane, together with hydrogen, makes the flatus combustible. Hydrogen sulfide is only a small fraction of a percent of the gas passed as flatus.

119

The amount of hydrogen and methane produced depends upon the nature of the unabsorbed residue. The hulls of beans contain oligosaccharides which cannot be hydrolyzed by intestinal oligosaccharidases, and when beans are eaten to the extent of 25% of the caloric intake, the rate of passage of flatus may be as high as 200 cc per hour. Deficiency of lactase leaves lactose to be fermented by intestinal flora, and when milk is drunk by a person with lactase deficiency, he makes as much as 4 cc of gas a minute. All varieties of fermentation produce organic acids that are partially neutralized in the small intestine or colon.

3. *Neutralization*. When 1 mEq of acid reacts with 1 mEq of bicarbonate at body temperature, 25 cc of carbon dioxide is liberated. During emptying of a meal from the stomach, 50 mEq of acid from the stomach may react with an equivalent amount of bicarbonate from the pancreas, and 1,250 cc of carbon dioxide bubbles out of duodenal contents. The partial pressure of carbon dioxide rises as high as 700 mm Hg, and most of the carbon dioxide resulting from neutralization in the small intestine diffuses back into the blood. Carbon dioxide resulting from reaction of organic acids formed in the colon with bicarbonate secreted by the colonic mucosa is more slowly absorbed, and after beans are eaten the mole fraction of carbon dioxide in flatus is as high as 0.62.

4. *Diffusion*. Addition of other gases to nitrogen in the gut reduces the partial pressure of nitrogen, and then nitrogen diffuses from blood into the gas phase in the gut at a rate of 1 to 2 cc a minute. Any oxygen diffusing from blood is used by colonic flora.

In man in the upright position, a gas bubble is trapped in the fundus of the stomach above the entrance of the esophagus. A characteristic volume for this bubble is 50 cc. There is another 100 cc of gas in transit through the small intestine and colon, making a total of 150 cc.

The stomach can be filled with a large volume of gas before belching, or eructation, occurs. When the gas bubble in the stomach is large enough to extend below the entrance of the esophagus, gas may escape into the esophagus when the lower esophageal sphincter relaxes, but it does not pass the hypopharyngeal sphincter. Distention of the esophagus by gas stimulates secondary peristalsis, and the gas is swept back into the stomach. This sequence of filling and emptying of the esophagus may repeat many times before eructation occurs. Then, when the esophagus is filled with gas, the jaw is thrust forward, the abdominal muscles contract, inspiration is attempted against a closed glottis and gas is expelled through the partially opened hypopharyngeal sphincter.

Gas moves rapidly through the intestinal tract, because it has low viscosity. A single segmental contraction in the small intestine or colon pushes a bubble of gas forward through several segments, whereas it pushes much more viscous chyme only a short distance. This difference in viscosity accounts for the fact that gas can be expelled through the anus without accompanying feces, for by the time the extremely viscous feces have moved, wind has been broken and the external anal sphincter has tightly contracted again. The external anal sphincter relaxes reflexly during micturition, and this accounts for the frequent passing of flatus while urinating.

It is not the fat content but the entrained gas bubbles that cause feces to float, and feces may be made to sink or rise like a Cartesian diver by increasing or decreasing the ambient pressure.

16. THE COLON

The parts of the colon and their names are shown in Fig. 16-1.

The intrinsic plexuses of the colon are similar to the plexuses of the small intestine, and they have a similar important role in regulating movements of the colon. The vagus nerve provides parasympathetic innervation to the proximal two-thirds of the colon. Parasympathetic innervation to the distal colon, the rectum and the internal anal sphincter comes from sacral segments of the spinal cord, and it is particularly important in governing defecation. Muscle at the anus is the external anal sphincter, which is composed of striated muscle, and therefore the behavior of the external anal sphincter is entirely controlled by motorneurons the cell bodies of which are in the sacral segments of the cord. Sympathetic innervation of the colon acts to constrict blood vessels, and it is partly excitatory, partly inhibitory for colonic smooth muscle.

In general, movements of the colon are very slow, and the first radiologist to observe the colon said that it presented a picture of still life. Slow movements of its contents are appropriate for the colon's function of absorbing water and electrolytes. It receives something like 500 ml of isotonic fluid a day from the ileum. The colonic mucosa actively reabsorbs Na^+ and Cl^-, and water is absorbed along with electrolytes. The colonic mucosa also secretes K^+ and HCO_3^- into its contents. Absorption of Na^+ and secretion of K^+ is partially controlled by aldosterone, and when aldosterone concentration in the plasma is elevated, absorption of Na^+ and secretion of K^+ are enhanced. Acids produced by fermentation in the colon are partially neutralized by secreted HCO_3^-. The absorptive capacity of the human colon is about 2,000 ml a day, and when

123

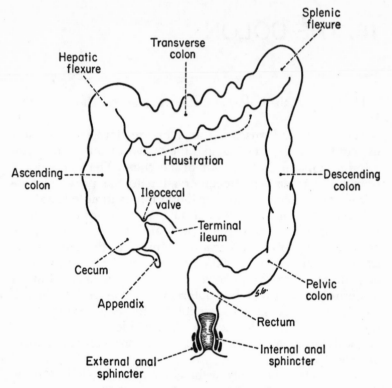

Fig. 16-1.—The parts of the colon.

this is exceeded by excessive delivery of fluid from the ileum, diarrhea results. Absorption of Na$^+$, and consequently of water, in the colon is inhibited by some bile acids, and therefore failure of absorption of bile acids in the terminal ileum can also be a cause of diarrhea.

Regularly spaced, ringlike contractions of the circular muscle of the colon divide it into *haustra*. Slow contraction of one ring of circular muscle is replaced by slow relaxation, and the neighboring, hitherto relaxed ring of circular muscle slowly contracts. This behavior of the circular muscle of the colon is similar to that causing segmentation in the small intestine, with the difference that colonic contractions and relaxations occur over many minutes, whereas those of the small intestine occur in seconds.

Much of the time the contents of the haustra are shuttled back and forth with no progressive movement. About one third of the time during the interdigestive period, haustral contents are moved in both directions, orthograde and retrograde with orthograde movement being slightly more frequent. Thus the contents are moved very slowly as absorption of water and electrolytes takes place.

Occasionally multihaustral segmental propulsion such as that shown in Figure 16-2 occurs. This type of movement displaces a substantial bulk of colonic contents in the orthograde direction.

Less frequently peristalsis consisting of a progressive wave of contraction preceded by a progressive wave of relaxation pushes colonic contents forward at a rate of 1 to 2 cm a minute. Reverse peristalsis may occur, but it is extremely rare in man.

In normal persons, between meals these several kinds of movements of the colon push the contents forward at the rate of about 8 cm an hour and backward at the rate of 3 cm an hour. Net forward movement is 5 cm an hour. In constipated persons, net forward movement is only 1 cm an hour. Slow and reluctant is the long descent with many a lingering farewell look behind.

After a meal, the colon is aroused from its torpor. Haustral shuttling to no effect decreases, and orthograde propulsion by segmentation increases. The frequency of multihaustral segmental propulsion doubles, and peristalsis is slightly more frequent. The result is that after eating, forward movement averages 14 cm an hour, backward movement 3 cm an hour and net forward movement 11 cm an hour. Administration of a long-lived cholinergic drug increases net forward movement to 20 cm an hour.

Increased motility of the colon after meals frequently pushes feces into the rectum and arouses the urge to defecate. Distention of the rectum is the primary stimulus for defecation. The basic mechanism of the defecation reflex re-

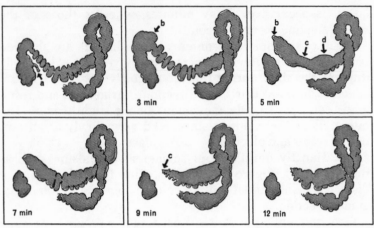

Fig. 16-2.—Multihaustral propulsion in the human colon. At the time of the first observation (*top left*), barium-impregnated ileal contents (*arrow a*) were entering the cecum. Contraction of the cecum, 3 minutes later, propelled a large part of its contents upward to distend the region of the hepatic flexure (*arrow b*). After another 2 minutes, this section had also contracted, and its contents were distributed over the proximal half of the transverse colon. Haustral markings disappeared over most of this length, and there was some narrowing of the mass in 2 to 3 inches of the middle section (*arrow c*). Material forced out of the narrow section was accommodated by distention of the next four haustra (*arrow d*). As haustration was returning, 2 minutes later, most of the proximal half of the transverse colon between *b* and *c* also contracted. All the contents expelled from this section lodged in the distal half of the transverse colon. The conical outline at *c* in the fifth picture is typical of multihaustral contraction. (Drawings made from cineradiographic studies; adapted from Ritchie, H. A.: Gut 9:442, 1968.)

sides in the intramural plexuses of the colon and the muscles of the rectum and internal anal sphincter. When the rectum is distended, its walls are stretched, and pressure in the rectum rises (Fig. 16-3). If the stretch is small or brief, there may be no reflex contraction of the rectum, and tension in the wall and pressure in the lumen may return to baseline levels. If stretch is greater, receptors in the wall of the rectum are stimulated, and, through a reflex arc confined to the intramural plexuses, the rectum is stimulated to contract. Pressure in the

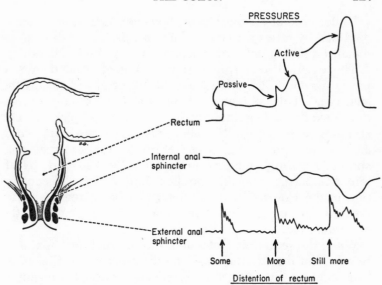

Fig. 16-3.—Response of the rectum, internal anal sphincter and external anal sphincter to distention of the rectum. Some distention of the rectum stretches its wall and causes a passive increase in pressure. More distention causes more passive increase in pressure and is followed by further increase in pressure resulting from active contraction. Still more distention is followed by greater active contraction, and subsequent contractions may occur rhythmically at about 20-second intervals. Each increase in pressure in the rectum is accompanied by a decrease in pressure in the internal anal sphincter and an increase in pressure in the external anal sphincter. (Rectal and internal anal sphincter pressures adapted from Denny-Brown, D., and Robertson, E. G.: Brain 58:256, 1935; external anal sphincter pressures adapted from Schuster, M. M., et al.: Bull. Johns Hopkins Hosp. 116:79, 1965.)

rectum then rises further. At the same time the internal anal sphincter, which is a thickening of the smooth muscle at the end of the rectum, relaxes. Contraction of the rectum and relaxation of the sphincter tend to expel rectal contents. This behavior takes place when there is no extrinsic innervation of the rectum and internal anal sphincter, and there is, therefore, no possibility that the defecation reflex can be modified through extrinsic reflexes.

⌈Under normal circumstances, however, this basic reflex is modified by reflex activity in the neuraxis. The reflex can be enhanced so that defecation occurs, or it can be inhibited.

The external anal sphincter is a ring of striated muscle partly surrounding the internal anal sphincter and closing the anal canal. This sphincter is kept in a state of tonic contraction by reflexes beginning in its own muscle spindles. The spindles send impulses to the sacral part of the spinal cord, and corresponding impulses return to the external anal sphincter through motorneurons controlled, in part, by afferent impulses from the spindles. The strength of contraction of the external anal sphincter can be increased or decreased by other reflex influences converging on the motorneurons.

When the rectum is distended, the internal anal sphincter relaxes, but the external anal sphincter contracts (Fig. 16-3). The stimulus for contraction is activation of stretch receptors in the wall of the rectum, the afferent fibers of which go to the spinal cord. This contraction of the external anal sphincter is a major factor providing for fecal continence.

Continence can usually be ensured by voluntary effort. When defecation is to be inhibited, impulses descend the spinal cord and, acting through the motorneurons to the external anal sphincter, cause it to close tightly. At the same time, impulses from the cord to the rectum and distal colon cause them to relax. Relaxation of the muscle of the rectum reduces tension in its wall, and although the rectum may be full of feces, stretch receptors are not activated. They no longer send afferent impulses to the intramural plexuses and to the spinal cord. There is now no urge to defecate, and there is no defecation reflex; defecation is postponed until more feces arrive to distend the rectum further.

Defecation is often inhibited by pain or fear of pain.

On the other hand, the defecation reflex can be facilitated. In this case, impulses arising at the highest level converge on a "defecation center" in the medulla, and from there facilitating impulses pass to the sacral section of the cord. The external anal sphincter and other perineal muscles relax, and the anal canal is partly extruded. Contraction of the longitudinal muscles of the rectum shortens the rectum and obliterates the angle the distal colon makes with the rectum. Peristaltic waves sweep from the sigmoid colon to the rectum, pushing feces through the relaxed internal and external anal sphincters at a rate ranging from stately slowness to explosive violence. At the end of defecation, there is rebound contraction of the external anal sphincter.

In a normal person, evacuation is assisted by a large increase in intra-abdominal pressure brought about by contractions of the chest muscles on a closed glottis and simultaneous contraction of abdominal muscles. The hemodynamic consequences of this maneuver—the Valsalva maneuver—are an abrupt rise in arterial pressure as the increase in intrathoracic pressure is transmitted across the wall of the heart, stoppage of venous return with subsequent fall in cardiac output and then a fall in arterial pressure. Death while straining at stool results from cerebrovascular accidents, ventricular fibrillation or coronary occlusion. Consequently, the appropriate place to look for an elderly person who does not answer the telephone is on the floor of the bathroom.

In some persons, particularly children, the ganglia of the intrinsic plexuses of a part of the distal colon or rectum degenerate for unknown reasons. Then the defecation reflex fails, and the colon above the affected part becomes enormously distended. This condition is called *megacolon.* If the aganglionic segment can be identified and removed, with anastomosis of the normal part of the colon above to the normal part below, the colon can empty regularly, and the distention disappears.

Psychosomatic factors, from the most deeply buried to the most overt, influence functions of the colon. The reaction of extreme fright and the use of the colon to express hostility or disdain are familiar to all. The irritable or spastic colon is characterized by abnormal motor function, abdominal pain and flatulence. Constipation alternates with diarrhea, and the condition waxes and wanes in step with changes in the life situation. Medical students need not be reminded of cycles of diarrhea in phase with examinations.

The normal natural frequency of defecation ranges from after every meal to once every 3 days or so. Consequently, it is difficult to specify what infrequency of defecation constitutes constipation. Nevertheless, when defecation is unduly postponed, the symptoms of constipation occur. These are mental depression, restlessness, dull headache, loss of appetite sometimes accompanied by nausea, a foul breath and coated tongue, and abdominal distention. This miserable state is exacerbated by fear of its consequences. Many persons are convinced that unless the colon is cleaned out regularly, toxins absorbed through the wall of the colon will poison them. Although some potentially toxic substances such as ammonia and cadaverine may be absorbed through the colonic mucosa into portal blood, in normal persons these are removed by the liver before they can reach the systemic circulation. Many of the symptoms appear to result from prolonged distention of the rectum, for they can be reproduced by stuffing the rectum with cotton and promptly alleviated by emptying the rectum.